Metabolic
Effects
of
Dietary Fructose

Authors

Sheldon Reiser, Ph.D.
Carbohydrate Nutrition Lab
U.S. Department of Agriculture
Beltsville Human Nutrition Research Center
Beltsville, Maryland

Judith Hallfrisch, Ph.D.
National Institutes of Health
National Institute on Aging
Gerontology Research Center
Baltimore, Maryland

CRC Press
Taylor & Francis Group
Boca Raton London New York

CRC Press is an imprint of the
Taylor & Francis Group, an **informa** business

First published 1987 by CRC Press
Taylor & Francis Group
6000 Broken Sound Parkway NW, Suite 300
Boca Raton, FL 33487-2742

Reissued 2018 by CRC Press

This book contains information obtained from authentic and highly regarded sources. Reasonable efforts have been made to publish reliable data and information, but the author and publisher cannot assume responsibility for the validity of all materials or the consequences of their use. The authors and publishers have attempted to trace the copyright holders of all material reproduced in this publication and apologize to copyright holders if permission to publish in this form has not been obtained. If any copyright material has not been acknowledged please write and let us know so we may rectify in any future reprint.

For permission to photocopy or use material electronically from this work, please access www.copyright.com (http://www.copyright.com/) or contact the Copyright Clearance Center, Inc. (CCC), 222 Rosewood Drive, Danvers, MA 01923, 978-750-8400. CCC is a not-for-profit organization that provides licenses and registration for a variety of users. For organizations that have been granted a photocopy license by the CCC, a separate system of payment has been arranged.

Trademark Notice: Product or corporate names may be trademarks or registered trademarks, and are used only for identification and explanation without intent to infringe.

Library of Congress Cataloging-in-Publication Data

Reiser, Sheldon, 1930-
  Metabolic effects of dietary fructose.

  Includes bibliographies and index.
  1. Fructose--Metabolism.   2. Fructose--Physiological
effect.   3. Fructose in human nutrition.   I. Hallfrisch,
Judith.   II. Title.   [DNLM: 1. Dietary Carbohydrates--
metabolism.   2. Fructose--metabolism. QU 75 R375m]
QP702.F7R45   1987   612'.396   86-31687
ISBN 0-8493-6457-4

A Library of Congress record exists under LC control number: 86031687

Publisher's Note
The publisher has gone to great lengths to ensure the quality of this reprint but points out that some imperfections in the original copies may be apparent.

Disclaimer
The publisher has made every effort to trace copyright holders and welcomes correspondence from those they have been unable to contact.

ISBN 13: 978-1-315-89531-4 (hbk)
ISBN 13: 978-1-351-07441-4 (ebk)

Visit the Taylor & Francis Web site at http://www.taylorandfrancis.com and the
CRC Press Web site at http://www.crcpress.com

PREFACE

There are a number of reasons why interest in the metabolic effects of dietary fructose has heightened in the past several years. With the introduction of high-fructose corn sweeteners in 1970, the amount of free fructose in the food supply of the U.S. has increased at least sixfold and fructose is rapidly becoming a major dietary ingredient. Several properties of fructose appear to be beneficial to segments of the population with metabolic diseases such as obesity and diabetes. Fructose is sweeter than sucrose and thus has potential as a weight loss diet sweetener. Fructose given alone does not appear to promote significant insulin secretion nor is its entry into tissues insulin dependent. However, there are other metabolic effects observed after feeding fructose to experimental animals and humans that appear to be undesirable. It will be the major purpose of the material to be presented here to focus on the effects observed after the feeding of fructose to experimental animals and humans, especially as this impacts on metabolites that are considered risk factors for diseases. Material pertaining to the interaction between dietary fructose and metabolic pathways not usually associated with carbohydrate metabolism such as those involving nucleotides, copper, amino acids, and ethanol will also be included. The influences of genetic predisposition and of interactions between fructose and other environmental factors on the magnitude of observed effects will be stressed and discussed as a possible source of apparently contradictory findings from studies using the same general experimental protocol. Mechanisms proposed to explain the metabolic effects of fructose will also be presented.

In many cases the metabolic effects of dietary sucrose appear to be due to its fructose moiety. It is therefore often difficult to differentiate between the metabolic effects produced by these sugars. Care will be taken not to use the effects obtained with sucrose as synonymous with those of fructose. However, it appears to be of nutritional importance to emphasize where the metabolic effects of fructose and sucrose are similar or disparate.

Each chapter in this book will be treated as an entity and will not depend on material presented in other sections for an understanding of its content. This format will necessarily produce a certain amount of overlapping and repetition of material, however, we feel it is desirable in order for the reader to obtain a complete understanding of each chapter without having to read the whole book.

It is hoped that the material presented in this book will provide the reader with a detailed description of the published research pertaining to the metabolic effects of dietary fructose, will define future research needs, and will stimulate interest in further research aimed at evaluating the advisability of the intake of fructose by humans.

We would like to express our appreciation to Drs. Meira Fields, June Kelsay, Charles Lewis, and Otho Michaelis IV for their help in reviewing the material in many of the chapters. We are especially grateful to Andrea Powell for her unfailing assistance, especially in times of stress, in organizing the material and supervising its completion.

THE AUTHORS

Sheldon Reiser is Research Leader of the Carbohydrate Nutrition Laboratory at the Beltsville Human Nutrition Research Center of the U.S. Department of Agriculture. He is also an Adjutant Professor in the Department of Food, Nutrition and Institution Administration at the University of Maryland. Dr. Reiser received a B.S. *cum laude* in Chemistry from the City College of New York in 1953 and was awarded a M.S. and Ph.D. in Biochemistry from the University of Wisconsin in 1957 and 1959, respectively. In 1960, Dr. Reiser accepted a joint appointment in the Departments of Biochemistry and Medicine at the Indiana University Medical Center in Indianapolis. His research there involved the mechanisms by which mammalian intestines transport amino acids and sugars. In 1973 he accepted his current position. His major research interest has been the effect of dietary carbohydrates on metabolic risk factors associated with disease states in various segments of the human population. Dr. Reiser has over 200 publications to his credit and is a member of the American Chemical Society, American Society of Biological Chemists, American Institute of Nutrition, American Society of Clinical Nutrition, Society for Experimental Biology and Medicine, and the New York Academy of Sciences.

Judith Hallfrisch is principle investigator of the gerontology nutrition study of the Baltimore Longitudinal Study of Aging. She is also Adjunct Professor at the University of Maryland and a visiting lecturer in nutritional biochemistry at the Johns Hopkins University Medical School. She received an A.B. in history from Indiana University and M.S. and Ph.D. degrees in Nutritional Sciences from the University of Maryland.

Dr. Hallfrisch worked as a Research Nutritionist in the Carbohydrate Nutrition Laboratory of the Beltsville Human Nutrition Research Center of the Agricultural Research Service until 1984. Her research there focused on effects of carbohydrates, especially fructose and sucrose on development and control of diabetes mellitus and cardiovascular disease in both animals and humans. In 1984 Dr. Hallfrisch accepted a position at the Gerontology Research Center of the National Institute on Aging where her research centers around nutritional factors affecting diseases of the elderly. She is a member of the American Chemical Society, American Association of Cereal Chemists, American Society for Clinical Nutrition, American Institute of Nutrition, Mid Atlantic Lipid Society, and the Gerontological Society of America.

TABLE OF CONTENTS

Chapter 6
Effects on Other Hormones
Judith Hallfrisch

Chapter 7
Lipogenesis and Blood Lipids
Sheldon Reiser

Chapter 8
Uric Acid and Lactic Acid
Sheldon Reiser

Chapter 9
Interaction with Other Nutrients
Sheldon Reiser

Chapter 1

FRUCTOSE OCCURRENCE IN FOODS AND PRODUCTION OF HIGH-FRUCTOSE CORN SWEETENERS

I. CHARACTERISTICS

Fructose, also called levulose and fruit sugar, is a naturally occurring hexose. It is found in berries and other fruits, and comprises approximately 50% of honey and sucrose. Dubrunfaut isolated pure fructose in 1847.[1] In 1874, fructose was reported to be better tolerated by diabetics than sucrose or glucose.[2] It has been used as a sweetener in special foods for diabetics since that time. β-D-Fructofuranose is the only form of fructose which is fermented by yeast. Fructopyranose is the only known crystalline form of fructose and is probably the sweetest naturally occurring sugar. Fructose is more soluble in water than is sucrose. A saturated sugar solution of 100 g at 25°C will contain 81 g of fructose, 67 g of sucrose, and 51 g of glucose.[3] It is the difference in this property, solubility, which makes sucrose the more popular sugar for use in certain products such as candies and frostings, and fructose more popular in beverages or as a sweetener in canned fruits. The high solubility of fructose makes it more hydroscopic and therefore candies or frostings which contain a high level of fructose may become more sticky or gooey than is desirable. Relative humidity changes of only 1% can alter the consistency of candy containing a high percentage of fructose.

Fructose has been reported to be 15 to 80% sweeter than sucrose,[4] but sweetness is dependent on pH, temperature, and concentration. A fructose solution is sweetest when it is dilute, cool, and slightly acidic, making it a useful sweetener in soft drinks.[5] In water, fructose can be detected at lower levels than sucrose; however, fructose in pear nectar was perceived as less sweet than sucrose.[6] Fructose cake was judged less sweet than sucrose cake, probably due to the browning reaction.[7]

II. OCCURRENCE

Fructose is the predominant monosaccharide in a number of fruits (Table 1) including apples, cherries, currants, and pears.[8] Fructose, as a component of sucrose, also occurs in considerable amounts in many fruits. According to U.S. Department of Agriculture information, simple carbohydrate from fruits provides 10 to 15% of total sugar intake.[9] From 1913 to 1969 the percentage of simple carbohydrate as fructose declined from 4 to 2% because the consumption of fresh apples, which are a good source of fructose, decreased from 60 to 16 lb per capita. During this same period there was an increase in consumption of citrus fruits, but since they contain less fructose than apples, total fructose intake was still lower. Fresh apple consumption in 1980 was approximately the same as in 1969. Fructose was reported to supply 7.4% of total simple carbohydrate intake of 214 g/day. Application of various herbicides was reported to alter simple sugar content of apples, blueberries, and peaches, but the differences in fructose were minimal.[10] Values for fructose in the edible portion of apples ranged from 6.5 to 8.3%, of blueberries from 5.4 to 6.1%, and of peaches from 0.6 to 0.7%. Sucrose content in the edible portion of these fruits ranged from 4.1 to 4.9% in apples, from 0.8 to 1.5% in blueberries, and from 6.1 to 6.8% in peaches.

There are a number of vegetables which contain measurable amounts of fructose, both free and as a component of sucrose (Table 2). Most vegetables contain 1 to 2% free fructose and up to 3% fructose as sucrose.[11] While this may not seem much, it may constitute a high percentage of the total number of calories of these vegetables. Inulin, which is a polymer of fructose (Figure 1), is present in chicory, sweet potatoes, and Jerusalem artichokes. Inulin

Table 1
AMOUNTS OF SUGARS IN FRUITS

Fruit	Fructose	Glucose	Sucrose
Apple[a]	5.0	1.7	3.1
Apricot[b]	1.3	1.7	5.8
Banana[a]	3.5	4.5	11.9
Blackberries[a]	2.9	3.2	0.2
Blueberries[b]	3.8	3.7	0.2
Cantaloupe[a]	0.9	1.2	4.4
Cherries[a]	7.2	4.7	0.1
Currants[a]	3.7	2.4	0.6
Dates[a]	23.9	24.9	0.3
Figs (dried)[a]	30.9	42.0	0.1
Gooseberries[a]	4.1	4.4	0.7
Grapes (white)[a]	8.0	8.1	—
Grapefruit[a]	1.2	2.0	2.9
Honeydew[b]	2.6	2.5	5.9
Orange[a]	1.8	2.5	4.6
Peach[a]	1.6	1.5	6.6
Pear (Bosc)[a]	6.5	2.6	1.7
Pineapple[a]	1.4	2.3	7.9
Plums (green gage)[a]	4.0	5.5	2.9
Prunes[a]	15.0	30.0	2.0
Raspberries[a]	2.4	2.3	1.0
Strawberries[a]	2.3	2.6	1.4
Tomatoes[a]	1.2	1.6	0.3
Watermelon[b]	3.5	1.8	2.3

[a]   Expressed as g/100 g edible portion. Adapted from Hardinge, M. G., Swarner, J. B., and Crooks, H., *J. Am. Diet. Assoc.*, 46, 197, 1965.

[b]   Expressed as percentage fresh basis. Adapted from Shallenberger, R. S., in *Sugars in Nutrition*, Sipple, H. L. and McNutt, K. W., Eds., Academic Press, Orlando, Fla., 1974, 67.

Table 2
AMOUNTS OF SUGARS IN VEGETABLES

Vegetable	(g/100 g fresh)		
	Fructose	Glucose	Sucrose
Asparagus	1.30	0.92	0.29
Beet	0.16	0.18	6.11
Cabbage	1.20	1.58	0.15
Carrot	0.85	0.85	4.24
Eggplant	1.53	1.51	0.25
Onion	1.09	2.07	0.89
Potato	1.15	1.04	1.69
Pumpkin	1.43	1.69	1.30
Radish	0.74	1.34	0.22
Squash, winter	1.16	0.96	1.61
Sweet corn	0.31	0.34	3.03
Sweet potato	0.30	0.33	3.37

Adapted from Shallenberger, R. S., in *Sugars in Nutrition*, Sipple, H. L. and McNutt, K. W., Eds., Academic Press, Orlando, Fla., 1974, 67.

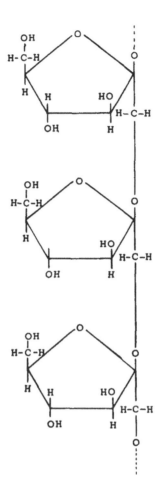

FIGURE 1. Structure of inulin,
a polymer of β-D-fructofuranose
linked by 2,1 bonds.

may be hydrolyzed to fructose at high temperatures under acid conditions. Fructose is also present in some legumes as the trisaccharide raffinose (Figure 2) and the tetrasaccharide stachyose (Figure 3). Raffinose contains one molecule each of galactose, glucose, and fructose. Stachyose contains two galactose molecules and one each of glucose and fructose. These compounds are not digested in the human small intestine. They are, however, fermented in the large intestine, and may be the source of flatulence which occurs after ingestion of legumes. A distribution of various sugars and these fructose-containing oligosaccharides occurring in legumes appears in Table 3.

Fructose occurs either free or as a component of sucrose in a variety of sweets (Table 4). Honey contains slightly more fructose than glucose, corn sugar is almost completely glucose, cane and maple sugar are predominantly sucrose, and molasses contains mostly sucrose, but also is about 8% fructose.

Although honey provides the highest concentration of fructose as a natural sweetener, relatively small amounts of it are consumed. Honey consumption per capita from the period 1960 to 1981 averaged about 1 lb/year with a range of 0.7 to 1.2 lb.[12] Edible syrup consumption (another minor source of fructose) during this same period declined 50% from only 0.8 lb per capita in 1960 to a steady 0.4 lb from 1974 to 1981. Beet sugar consumption ranged from 24.5 to 32.0 lb per capita.

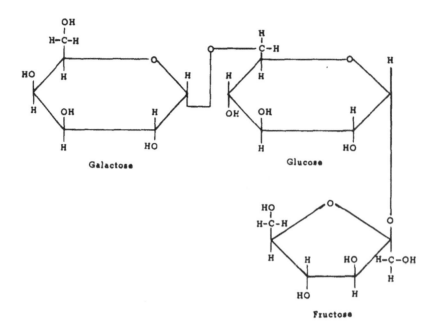

FIGURE 2.   Structure of raffinose, α-D-galactopyranosyl-(1,6)-α-D-glucopyranosyl-(1,2)-β-D-fructofuranoside.

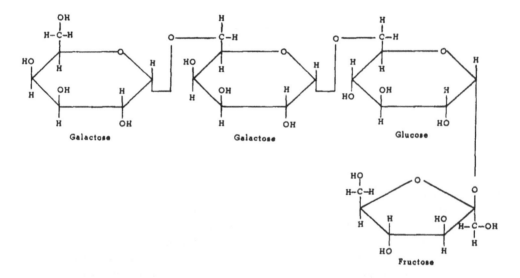

FIGURE 3.   Structure of stachyose, α-D-galactopyranosyl-(1,6)-α-D-galactopyranosyl-(1,6)-α-D-glucopyranosyl-(1,2)-β-D-fructofuranoside.

III. HIGH-FRUCTOSE CORN SWEETENERS

A. Production

The two major sources of fructose in the U.S. diet are cane sugar and high-fructose corn sweeteners. Cane sugar consumption averaged approximately 100 lb per capita per year from 1960 to 1974.[9] Cane sugar is produced from stalks of sugar cane from which the leaves have been removed. Heavy rollers crush these stalks to express the juice.[13] The 10 to 15% sugar juice is treated with lime and filtered to remove impurities. This filtered solution is then evaporated until concentrated enough for crystallization. The crystals that are formed

Table 3
**SUGARS AND OLIGOSACCHARIDES CONTAINING
FRUCTOSE IN VARIOUS LEGUMES**

	(g/100 g weight)				
Legume	Fructose	Glucose	Sucrose	Raffinose	Stachyose
Fava bean	0.18	—[a]	3.36	0.66	—
Lima bean	0.08	0.04	2.59	0.20	0.59
Snap bean	1.20	1.08	0.25	0.11	0.19
Alaska pea	0.08	—	3.00	0.06	0.06
Wrinkled pea	0.23	0.32	5.27	0.58	0.49
Cow pea	0.06	0.08	1.86	0.10	1.66
Dry bean	—	—	2.40	0.80	3.40
Mung bean	—	—	1.19	0.40	1.75
Pea bean	—	—	2.55	0.65	3.06
Pea seed	—	0.24	4.11	1.75	7.96
Soybean	—	—	4.53	0.73	2.73

[a]   Dashes indicate no available data. Adapted from Shellenberger, R. S., in *Sugars in Nutrition*, Sipple, H. L. and McNutt, K. W., Eds., Academic Press, Orlando, Fla., 1974, 67.

Table 4
AMOUNTS OF SUGARS IN SYRUPS AND SWEETS

	(g/100 g edible portion)		
Syrups and sweets	Fructose	Glucose	Sucrose
Chocolate, sweet dry	—[a]	—	56.4
Corn sugar	—	87.5	—
Corn syrup (medium conversion)	—	47.0[b]	—
Honey	40.5	34.2	1.9
Royal jelly	11.3	9.8	0.9
Maple syrup	—	—	62.9
Molasses	8.0	8.8	53.6
Cane sugar	0	0	99.5

[a]   Dashes indicate lack of acceptable data.
[b]   Includes glucose from maltose.

Adapted from Hardinge, M. G., Swarner, J. B., and Crooks, H., *J. Am. Diet. Assoc.*, 46, 197, 1965.

are covered with molasses. The removal of the molasses and further purification to produce white granulated sugar is called "refining". Crystals are washed, centrifuged to remove the molasses, and then dissolved in warm water, and most remaining impurities are removed by filtration or precipitation. Charcoal is used to remove the last remaining impurities. New crystals are then formed in another evaporation under a vaccuum at low temperature, and are separated according to size. Beet sugar is a minor source of sucrose made by the same general process, except that in the first step sugar is extracted with hot water rather than by crushing.

The introduction of high-fructose corn sweeteners in 1967 has led to their steady replacement of cane sugar in many processed foods.[13] In 1981 cane sugar consumption was less than 80 lb per capita and high-fructose corn sweeteners had increased to about 20 lb.

There are three major high-fructose corn sweeteners used in processed foods in this country.

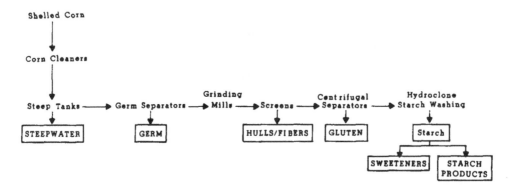

FIGURE 4.   Production of corn syrups by a corn wet milling process. (Adapted from Nutritive Sweetness from Corn, Corn Refiners Association, Inc., Washington, D.C., 1984.)

They contain 42, 55, or about 90% fructose. The first mass-produced high-fructose corn sweeteners contained only about 15% fructose, with the remainder being dextrins or more complex glucose polymers. In 1968, 42% fructose was produced in a batch process and in 1973 in a continuous process. In 1977, high-fructose corn sweetener containing 55% fructose was introduced by combining 90% fructose with 42% fructose corn sweetener.

The discovery that starch was a polymer of glucose and could be transformed to a sweetener by heating with acid was an accident. G. S. C. Kirchoff, a chemist in a Russian ceramics laboratory, was looking for a substitute for gum arabic.[13] Corn sugar was first produced commercially in 1866 in this country at a plant in Buffalo, N.Y. The process by which corn syrups are produced is shown in Figure 4. Corn is washed and then soaked or "steeped" in warm water containing sulfur dioxide. This softens and swells the corn, facilitating subsequent separations. The steeped corn is degerminated and then ground. The hulls and fibers are removed by passing the ground slurry through screens. Gluten is then separated by centrifugation. The remaining starch fraction is then washed, separated into fractions for starch products or conversion to syrups.

The conversion of starch to glucose can be catalyzed by either acid or enzymes. Products resulting from hydrolysis include fucose, maltose, and maltotriose, as well as longer dextrins. β-Amylase hydrolyzes the starch chain at every other bond, producing maltose. α-Amylase hydrolysis produces longer chains of soluble saccharides. Glucoamylase hydrolyzes these polymers unit by unit, resulting in a dextrose hydrolysate. The differences in composition between glucose and fructose are slight; only the configurations around the first and second carbons are different (Figure 5). Fructose is a ketosugar while glucose is an aldehyde. Sweetness is related to both the nature of the reducing group and the spatial arrangement of the atoms.[14] A number of theories have been proposed to explain differences in sweetness of various sugars by differences in molecular geometry.[13] Hydrogen bonding was proposed to be the cause of sweetness in certain sugars where the negative atom is bonded to hydrogen, acting as an acid, and another close negative atom acts as a base. Fructose has a relative sweetness of 120 to 160, while sucrose is 100 and glucose has only a sweetness of 70 to 80. Therefore, conversion of the glucose from corn sugars to fructose substantially increases sweetness.

The enzymatic conversion of glucose to fructose was first introduced commercially in 1967. Initially the isomerase was added directly to glucose syrups produced by the wet milling process. Now, however, a continuous process has replaced this batch production. The isomerase is now immobilized by an insoluble carrier. The higher-level fructose corn syrups (90%) are now possible due to the development of a fractionation process which separates newly formed fructose from the glucose. The solution containing glucose plus isomerized fructose is passed through a column.[13] The fructose adheres to divalent calcium

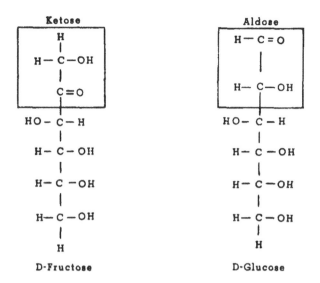

FIGURE 5. Differences in the structures of fructose and glucose.

Table 5
PAST AND PROJECTED SUGAR AND
HIGH-FRUCTOSE CORN SWEETENERS
AS A PERCENTAGE OF TOTAL
CALORIC SWEETENERS CONSUMED
DURING THE INDICATED YEAR[a]

Year	Refined sugar	High-fructose corn sweeteners
1970	83.6	0.6
1975	75.5	4.2
1980	66.9	15.3
1985	51.5	30.4
1990	49.3	32.2

[a] Remaining percent from glucose and minor caloric sweeteners such as syrup and honey.

Adapted from Barry, R. D., High-Fructose Corn Syrup and Pure Fructose: Potential Consumption and Substitution for Sugar by 1990, AMA Resour. Conf. HFCS and Fructose, Economic Research Service, U.S. Department of Agriculture, Washington, D.C., 1984.

in the column; the glucose passes through and is then recirculated for isomerization. The fructose is then flushed from the column with deionized water. Fructose concentration of this solution is 80 to 90%.

B. Consumption

High-fructose corn sweeteners are replacing sucrose as the major sweetener in a number of processed foods. In 1981 cane sugar consumption had decreased to less than 80 lb per capita and high-fructose corn sweeteners had increased to about 20 lb.[15] Table 5 shows the decline of sucrose consumption and the increase of high-fructose corn sweeteners as a percentage of total caloric sweeteners. By 1990 cane sugar is projected to provide less than 50% of total caloric sweeteners. Projected use of sugar and high-fructose corn sweeteners

Table 6
PAST AND PROJECTED PER CAPITA SWEETENER
CONSUMPTION BY TYPE OF USE

	High-fructose corn sweeteners (lb per capita)			Cane sugar		
	1970	1980	1990	1970	1980	1990
Beverages	—	9.4	25.6	23.0	19.0	3.0
Processed food	0.3	4.1	6.2	9.4	4.7	3.6
Cereal and bakery products	0.4	3.3	3.9	13.8	11.7	12.0
Dairy	0.1	1.2	2.3	5.2	4.0	3.0
Confections	—	0.2	0.4	10.7	8.2	9.0
Other	—	1.0	1.0	5.0	6.2	4.5
Total industrial	0.7	19.2	39.4	67.2	53.7	38.1
Nonindustrial	—	—	—	36.3	29.4	26.0

Adapted from Barry, R. D., High-Fructose Corn Syrup and Pure Fructose: Potential Consumption and Substitution for Sugar by 1990, AMA Resour. Conf. HFCS and Fructose, Economic Research Service, U.S. Department of Agriculture, Washington, D.C., 1984.

in processed foods is shown in Table 6. One of the major uses of high-fructose corn sweeteners is in soft drinks.[15] The only sweetener now in fountain syrup Coca-Cola® is 55% high-fructose corn sweetener which provides 75% of the sweetener in bottled and canned Coke®. Pepsi® fountain syrup is 80% high-fructose corn sweetener-55, and about 50% of packaged Pepsi® now contains high-fructose corn sweeteners.

High-fructose corn sweeteners in beverages are projected to rise from 9.4 lb per capita in 1980 to 25.6 lb in 1990. Sucrose use in soft drinks is expected to drop precipitously during this same period. High-fructose corn sweetener use in processed foods does not constitute a major source of fructose, but is expected to increase to about 6 lb per capita by 1990 while sugar intake from this source decreases slightly.

Dairy products provide only a minor amount of fructose, both from high-fructose corn sweeteners and sucrose. Confections and other uses provide about 7 lb per capita fructose, mostly from sucrose. Nonindustrial use of high-fructose corn sweetener is negligible, but sucrose provides about 13 lb per capita per year of fructose as home table sweetener and in homebaked and canned goods.

Due to physical and chemical properties of fructose corn sweeteners in baking,[16] high-fructose corn sweeteners are not a major source of fructose in these foods. Sucrose, however, does provide about 6 lb per capita of fructose from cereal and bakery goods. Gas chromatography has been used to analyze various sugars in breakfast cereals; fructose ranged from not detectable to 9.5%, but sucrose was <0.1 to 55.0% of dry weight.[17] Table 7 lists the sugar content of some cereals which were analyzed by gas chromatography.

Crystalline fructose is prepared from glucose sources in an isomerization process as are the corn sweeteners. It is then evaporated and crystals formed in much the same way that cane sugar is prepared. Crystalline fructose is produced for human consumption in the U.S. by Hoffman-LaRoche, Nutley, N.J. It is much more expensive than either corn sweeteners or cane sugar. Although crystalline fructose is widely used in Europe, total use in this country is only about 10,000 ton/year.[15] It is sold as a specialty sweetener in many health food stores and at this time does not constitute a major source of fructose consumption.

Table 7
AVERAGE SUGAR CONTENT OF SELECTED READY-TO-EAT CEREALS DETERMINED BY GAS CHROMATOGRAPHY

Cereal	Fructose	Glucose	Sucrose	Total sugars
		(% dry weight)		
Puffed Rice	ND	ND	<0.1	<0.1
Puffed Wheat	0.1—0.3	ND	0.1—0.4	<0.5
Shredded Wheat	ND	ND	0.64—0.78	0.64—0.78
Cheerios	0.06—0.14	ND	2.58—3.48	2.58—3.48
Corn Chex	ND	ND	3.45—3.79	3.45—3.79
Wheat Chex	0.68—0.84	0.51—0.56	2.06—2.11	3.51—3.79
Rice Chex	0.15	0.16—0.20	3.95—4.23	4.28—4.58
Special K	0.24—0.30	0.22—0.24	4.85—4.99	4.85—4.99
Corn Flakes	1.08—1.18	1.16—1.58	2.23—3.33	5.39—6.05
Product 19	0.84—0.94	0.82—0.91	7.84—8.36	9.53—10.0
Bran Flakes, 40%	0.85—1.32	0.76—1.17	9.22—10.7	11.6—14.4
Life	ND	ND	15.3—17.7	15.3—17.7
All Bran	1.09—1.64	0.90—0.98	15.6—16.0	19.3—19.4
C. W. Post (raisin)	3.84—5.48	4.28—5.90	18.1—18.5	27.4—31.1
Raisin Bran	9.36—9.51	7.87—8.23	11.4—11.6	28.7—29.4
Alphabits	<0.5	<0.5	35.4—39.9	35.4—39.9
Super Sugar Crisp	2.20—2.40	4.49—4.93	35.0—36.6	44.9—45.8
Apple Jacks	0.21—0.32	0.27—0.42	52.4—55.6	52.4—55.6
Sugar Smacks	1.23—1.34	11.3—11.8	41.4—44.0	53.9—57.1

Note: ND — not detectable. The detection limit was 0.01% for fructose and glucose, and 0.1% for sucrose.

Adapted from Li, B. W. and Schuhmann, P. J., *J. Food Sci.*, 45, 138, 1980.

IV. SUMMARY

Fructose occurs naturally as a component of fruits and vegetables. Honey is a mixture of approximately one half fructose and one half glucose. Fructose also occurs naturally as a component of a number of carbohydrates. The most predominant of these is the disaccharide sucrose, which is a combination of fructose and glucose. Many fruits, vegetables, and natural sweeteners and syrups contain considerable amounts of sucrose. Legumes contain the fructose-containing oligosaccharides, raffinose and stachyose. The major sources of fructose in the diet of people living in the U.S. are sucrose and high-fructose corn sweeteners. These sweeteners contain 42, 55, and 90% fructose and are produced from glucose corn sweeteners by an isomerization process. Since high-fructose corn sweeteners were introduced in 1967, the amount of fructose consumed as the monosaccharide in the American diet has increased spectacularly. High-fructose corn sweeteners are replacing sucrose in beverages and processed foods, but due to the higher solubility and hydroscopic properties of fructose, sucrose continues to be used as the major sweetener in baked goods and confections, as well as in industrial use. Crystalline fructose is produced from glucose by isomerization, but at this time its high cost precludes its becoming a major source of fructose in the diet.

REFERENCES

1. **Doty, T.,** Fructose: the rationale for traditional and modern uses, in *Carbohydrate Sweeteners in Foods and Nutrition,* Koivistoinen, P. and Hyvonen, L., Eds., Academic Press, London, 1980, 259.
2. **Kulz, E.,** *Beitrag zur Pathologie und Therapie des Diabetes Mellitus,* Elwert's Verlag, Marburg, 1874.
3. **Sertzen, G. and Linqvist, G.,** Comparative evaluation of carbohydrate sweeteners, in *Carbohydrate Sweeteners in Foods and Nutrition,* Koivistoinen, P. and Hyvonen, L., Eds., Academic Press, London, 1980, 127.
4. **Shallenberger, R. S.,** Hydrogen bonding and the varying sweetness of the sugars, *J. Food Sci.,* 28, 584, 1963.
5. **Prescott, R.,** Per capita food consumption highlights, *Natl. Food Rev.,* Fall, 7, 1981.
6. **Pangborn, R. M.,** A critical analysis of sensory responses to sweetness, in *Carbohydrate Sweeteners in Foods and Nutrition,* Koivistoinen, P. and Hyvonen, L., Eds., Academic Press, London, 1980, 87.
7. **Koivistoinen, P. and Hyvonen, L.,** Modifications of sweetness by food environment, in *Carbohydrate Sweeteners in Foods and Nutrition,* Koivistoinen, P. and Hyvonen, L., Eds., Academic Press, London, 1980, 163.
8. **Hardinge, M. G., Swarner, J. B., and Crooks, H.,** Carbohydrate in foods, *J. Am. Diet. Assoc.,* 46, 197, 1965.
9. **Wotecki, C. E., Welch, S. O., Raper, N., and Marston, R. M.,** Recent trends and levels of dietary sugars and other caloric sweeteners, in *Metabolic Effect of Utilizable Dietary Carbohydrates,* Reiser, S., Ed., Marcel Dekker, New York, 1982, 1.
10. **Mason, B. S., Eheart, J. F., and Welker, W. V., Jr.,** Total solids and sugars in herbicide-treated fruits, *J. Am. Diet. Assoc.,* 55, 562, 1969.
11. **Shallenberger, R. S.,** Occurrence of various sugars in foods, in *Sugars in Nutrition,* Sipple, H. L. and McNutt, K. W., Eds., Academic Press, Orlando, Fla., 1974, 67.
12. Economic Research Service, Sugar and Sweetener Outlook and Situation Report No. SSRV 7N2, U.S. Department of Agriculture, Washington, D.C., 1982, 21.
13. Nutritive Sweetness from Corn, Corn Refiners Association, Inc., Washington, D.C., 1984.
14. **Birch, G. G.,** Theory of sweetness, in *Carbohydrate Sweeteners in Foods and Nutrition,* Koivistoinen, P. and Hyvonen, L., Eds., Academic Press, London, 1980, 61.
15. **Barry, R. D.,** High-Fructose Corn Syrup and Pure Fructose: Potential Consumption and Substitution for Sugar by 1990, AMA Resour. Conf. HFCS and Fructose, Economic Research Service, U.S. Department of Agriculture, Washington, D.C., 1984.
16. **Doescher, L. C. and Hoseney, R. C.,** Effect of sugar type and flour moisture on surface cracking of sugar-snap cookies, *Cereal Chem.,* 62, 263, 1985.
17. **Li, B. W. and Schuhmann, P. J.,** Gas-liquid chromatographic analysis of sugars in ready-to-eat breakfast cereals, *J. Food Sci.,* 45, 138, 1980.

Chapter 2

INTESTINAL DIGESTION OF SUCROSE AND ABSORPTION OF FRUCTOSE

I. INTRODUCTION

The major sources of dietary fructose in the U.S. are sucrose and high-fructose corn sweeteners. In 1984, the per capita consumption of sucrose and high-fructose corn sweeteners in this country was estimated to be 67.5 and 36.3 lb/year, respectively.[1] From these estimates it can be calculated that the daily consumption of fructose from these sources is 64.5 g: 42 g from sucrose and 22.5 g from high-fructose corn sweeteners (average fructose content of 55%). In addition, fructose may enter the diet in the form of the indigestible polysaccharide inulin which is comprised of fructose units linked to each other in a linear fashion by β 2,1 bonds.

The fundamental unit of carbohydrate absorption is generally believed to be the monosaccharide. A major function of the small intestine is to hydrolyze those components of the carbohydrate diet that are not absorbable as such into their constituent monosaccharide units. This digestive function is carried out by enzymes that are either secreted into the duodenum from the pancreas or associated with the brush border membrane of the intestinal epithelial cell. The other major function of the small intestine is to pass the products of digestion into the blood or lymph. This is accomplished by specific absorption sites located in the brush border of the epithelial cell. In this chapter, the processes mediating the ultimate entrance of the various forms of dietary fructose into the portal blood will be described and the adaptability of these processes to different nutritional conditions will be examined.

II. DIGESTION OF SUCROSE

A. Mechanism

Of the sources of dietary fructose, only sucrose and inulin require prior digestion before the fructose moiety is ready for absorption. None of the digestive enzymes appear to be able to hydrolyze inulin into fructose. The resistance of inulin to hydrolysis is the basis for its use in the measurement of extracellular space and renal clearance.

The digestion of sucrose is mediated by enzymes located in the brush border membrane of the intestinal epithelial cell. Specifically, these disaccharidases have been shown to be present in mushroom-shaped particles, 50 to 60 Å in diameter, located at the external surface of the brush border membrane.[2,3] Figure 1 shows an electron micrograph of these particles as they appear on negatively stained brush border preparations isolated from hamster intestine. The location of the hydrolytic enzymes in the brush border appears to be external to the site of monosaccharide transport, suggesting a functional organization of the digestive and absorptive processes in the epithelial cell membrane. Two of the many disaccharidases located in the brush border are associated with both sucrase and isomaltase activities.[4,5] In the normal intestine, the sucrase-isomaltase complex is synthesized as a single chain precursor protein called pro-sucrase-isomaltase.[6] Synthesis occurs all along the intestinal villus but is maximal at the crypt-villus junction where mucosal cells are most actively undergoing differentiation.[7] After arriving at the brush border membrane, the pro-sucrase-isomaltase is cleaved into its active subunits by pancreatic proteases.[8] At the brush border surface, sucrase catalyzes the hydrolysis of sucrose to fructose and glucose.

B. Dietary Adaptation

The process of intestinal digestion shows a remarkable capacity to adapt to changes in

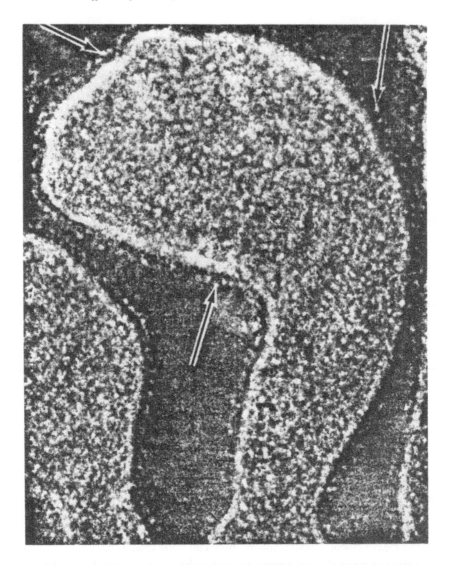

FIGURE 1. The intact, isolated brush border microvillus. Arrows identify the particles as projections along the edge or as globules when viewed head-on. Negatively stained with 2% potassium phosphotungstate. (Magnification approximately × 96,000.) (Reprinted from Johnson, C. F., *Fed. Proc., Fed. Am. Soc. Exp. Biol.*, 28, 26, 1969. Copyright 1969 by the AAAS. With permission.)

caloric and nutritional composition, feeding patterns, and the nutritional needs of the organism. Since the process of adaptation permits the organism to survive under a variety of environmental changes, digestive adaptation probably encompasses a broad spectrum of changes. Some changes are rapid, of short duration, and insure the survival of the organism for short periods of time. Other changes are chronic, more gradual, and permit the organism to adapt to prolonged stimulation by establishing a new steady state. This section will focus on adaptive changes in the activity of intestinal sucrase observed in experimental animals and humans due to changes in dietary composition.

Numerous studies have now established that the amount and nature of the dietary carbohydrate can influence markedly the activity of sucrase in the rat.[7,9-19] In general, sucrase activity is significantly increased when the rat is fed a high- as compared to a low-carbohydrate diet.[9,10,12,16-18] However, studies on the comparative effects of different sources of

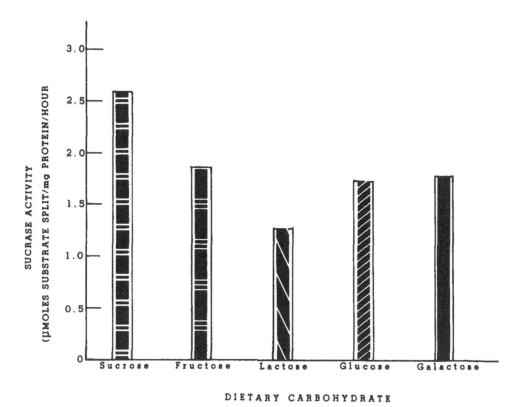

FIGURE 2. Effect of feeding diets containing 70% sucrose, fructose, lactose, glucose, or galactose for 24 hr on sucrase activity in rats previously fed a chow diet for 11 to 12 weeks. Each value is the mean from five rats. (Adapted from Koldovsky, O., Bustamante, S., and Yamada, K., *Mechanisms of Intestinal Adaptation*, Robinson, J. W. L., Dowling, R. H., and Riecken, E.-O., Eds., MTP Press, Hingham, England, 1982, chap. 12.)

mono- and disaccharides on sucrase activity have not provided such consistent results. Rats starved for 3 days and then refed a diet containing 68% sucrose for 24 hr had significantly greater levels of sucrase activity than did rats refed a diet containing 68% maltose, but not rats refed diets containing 68% fructose or glucose.[10] In contrast, sucrase activity in rats fed diets containing 70% carbohydrate as sucrose for 60 days was significantly greater than sucrase activity in rats fed either 70% glucose or lactose, but not in rats fed 70% maltose.[11] Figure 2 shows the effect on sucrase activity after the feeding of diets containing 70% sucrose, fructose, lactose, glucose, or galactose for 24 hr to rats previously fed a chow diet for 11 to 12 weeks.[17] Although sucrase activity was at least 39% higher after the rats consumed sucrose as compared to any of the other mono- or disaccharides, this difference was significant ($p < 0.05$) only for lactose and galactose.

Starch appears to be as effective as sucrose in adaptively increasing sucrase activity in rat intestine. Rats fed diets containing 70% of calories from either sucrose or starch for 3 to 44 days generally showed the same magnitude of increase in intestinal sucrase activity as rats fed a low-carbohydrate, high-fat diet.[16] Table 1 presents the sucrase activity in the proximal and middle intestine after feeding these diets for 44 days. In another study,[7] sucrase activity of rats fed diets containing either 65% sucrose or starch for 4 days was not found to be significantly different, but was significantly higher than sucrase activity in rats fed diets either lower in carbohydrate (30%) carbohydrate-free. These results may be due to an equal induction of the sucrase-isomaltase complex either by sucrose or by isomaltose liberated from the partial hydrolysis of the amylopectin component of dietary starch.

Sucrase levels in the small intestine of humans also show dietary adaptations. Feeding

Table 1
SUCRASE ACTIVITY IN THE PROXIMAL
AND MIDDLE PORTION OF THE SMALL
INTESTINE AFTER MALE RATS WERE FED
SUCROSE- OR STARCH-CONTAINING DIETS
FOR 44 DAYS

Diet (% calories)	n	Sucrase activity (μmol substrate hydrolyzed per milligram protein per hour)	
		Proximal intestine	Middle intestine
5% Sucrose	13	1.7 ± 0.38[a]	0.57 ± 0.01[a]
70% Sucrose	15	3.1 ± 0.28[b]	2.40 ± 0.36[b]
70% Starch	15	3.3 ± 0.44[b]	2.61 ± 0.26[b]

Note: Each value represents the mean ± SEM from the number of animals indicated.

[a,b] Values within a column with different superscript letters are significantly different from each other as determined by a one-way analysis of variance.

Adapted from Bustamante, S., Gasparo, M., Kendall, K., Coates, P., Brown, S., Somawane, B., and Koldovsky, O., *J. Nutr.*, 111, 943, 1981.

normal subjects diets containing 40 to 80% of calories as sucrose as compared to glucose for 7 to 20 days increased the activity of sucrase.[20] The feeding of fructose as compared to glucose in two subjects produced increases in sucrase activity similar to that noted with sucrose. Maltase activity was also increased by sucrose feeding, an effect probably due to the maltolytic activity of the sucrase-isomaltase complex.[21] Sucrose feeding had no effect on intestinal lactase activity. These results indicate that humans consuming diets containing high levels of sucrose and/or fructose can hydrolyze dietary sucrose very efficiently. The increase in sucrase activity occurred 2 to 5 days after subjects were changed from a glucose to a sucrose diet.[22] Since this response time is similar to the estimated time for intestinal cell turnover, it was suggested that this dietary adaptation occurred at the crypt cell level. It has also been reported that the synthesis or catabolism of disaccharidases may be modified in the mature villus cells.[23]

The mechanisms responsible for brush border enzyme adaptation are incompletely understood. Earlier studies used starvation to reduce disaccharidases to their lowest levels and then the animals were fed diets containing various amounts and types of carbohydrate to observe the responses to diet of certain of these enzymes. More recent studies have replaced the starvation pretreatment with isocaloric, carbohydrate-free diets or ones composed of nondigestible carbohydrates in order to lower disaccharidase activities.[19-24] Comparative studies of the two methods of lowering membrane disaccharidase activities indicate that the longer response time seen in starvation studies was a result of a generalized mucosal atrophy and not due to sluggishness of the intestinal adaptive response.[7,19,25] It appears that dietary adaptive changes in disaccharidase activities occur very rapidly in rats that have not been starved. In rats changed from a diet providing 70% of calories as starch to one which provided 5% of the calories as starch, there was a decrease in sucrase, maltase, lactase, and glucoamylase activities within 24 hr.[25]

A recent study has used a double labeling technique to estimate rates of sucrase synthesis and degradation.[7] When a diet containing 65% sucrose was fed to rats, sucrase activity was increased above control levels due to a threefold increase in sucrase synthesis and a slight decline in sucrase degradation. These results are consistent with the finding that the increased sucrase activity following the feeding of sucrose was the result of accumulation of newly synthesized enzyme.[26] One way that dietary carbohydrates can affect enzyme synthesis is through transcription. Starvation results in a decline in intestinal chromatin-bound RNA polymerase I, but no significant change in chromatin-bound RNA polymerase II when compared to the fed rat.[27] Feeding the starved rat a 70% sucrose solution produced a rapid and significant increase in both chromatin-bound RNA polymerases I and II. Although chromatin-bound RNA polymerase I values did not reach those of the nourished rat, RNA polymerase II values exceeded the nourished values 15 hr after sucrose feeding. These results indicate that the intestinal adaptive response to dietary sucrose involves modulation in gene transcription.

Hormones are frequently implicated in the intestinal adaptive response. Serum insulin levels drop during starvation and insulin has been reported to be necessary in the maturation of disaccharidases.[28,29] However, insulin does not appear to be necessary for the dietary adaptation of intestinal disaccharidases. Starvation for 5 days produced a decreased concentration of sucrase and maltase.[30] Serum insulin and glucose levels were also significantly reduced. If a 20% glucose solution was made available during the fast, the sucrase and maltase levels remained normal in the duodenum and increased in the jejunum. Intramuscular injection of insulin during the fast did not maintain sucrase levels. If both insulin and glucose were provided during the fast, the presence of insulin did not increase the levels of the enzymes above that observed with glucose alone.

III. ABSORPTION OF FRUCTOSE

A. Mechanism

The process by which fructose liberated from the digestion of sucrose and fructose derived directly from dietary sources is absorbed by the small intestine is based on studies using various in vitro intestinal preparations from experimental animals, primarily the rat. Fructose absorption was long considered to be a facilitated passive process.[31] However, more recent findings indicate that fructose absorption, at least in rat intestine, is an active process mediated by a specific transport carrier.

Fructose has long been known to be absorbed at rates intermediate between those of the actively transported monosaccharides such as glucose and galactose and those monosaccharides transported by passive diffusion.[32] The greater rate of fructose absorption as compared with passively transported monosaccharides was ascribed to the ability of the intestine to metabolize fructose to glucose. However, subsequent findings have shown that the conversion of fructose to glucose occurs only to a very limited extent in the small intestine of the rat,[33-35] chicken,[36] and man.[37,38] By the criterion of transport against a concentration gradient, there is evidence that fructose transport in rat intestine is an active process.[39-42] It now appears that fructose transport in mammalian intestine is mediated by a relatively specific carrier mechanism with properties distinct from those of the carrier utilized by other monosaccharides.[39,40,43,46] Table 2 summarizes some of the properties of this fructose transport system. In general, fructose transport has been found to be a Na^+-independent process that is not inhibited by phlorizin or glucose and is only slightly inhibited by the structurally similar ketose L-sorbose. Figure 3 presents a hypothetical model for fructose transport in rat small intestine based on these properties.

B. Dietary Adaptation

The independent transport pathways mediating the intestinal absorption of monosaccharides such as fructose and glucose appear to be influenced by the nature of the dietary

Table 2
SOME PROPERTIES OF FRUCTOSE ABSORPTION IN MAMMALIAN INTESTINE

Mammal	Intestinal preparation	Active transport	% Inhibition by				Na$^+$-dependent	Ref.
			D-Glucose	L-Sorbose	Phlorizin	Dinitrophenol		
Rat	Segments	Yes	24	31	13—51	33—47	Yes	39
	Segments	Yes	32	10	None	60	Possibly	40
	Cells	Yes	—	—	—	—	—	41
	Cells	Yes	—	—	—	—	—	42
Rabbit	Segments	—	None	14	None	—	No	43
Rat	Rings	No	None	None	None	—	No	44
Hamster	Segments	—	None	None	None	—	No	45
Rat	Brush border membranes	—	None	None	None	—	No	46
	Cells	Yes	7	—	—	—	—	47

17

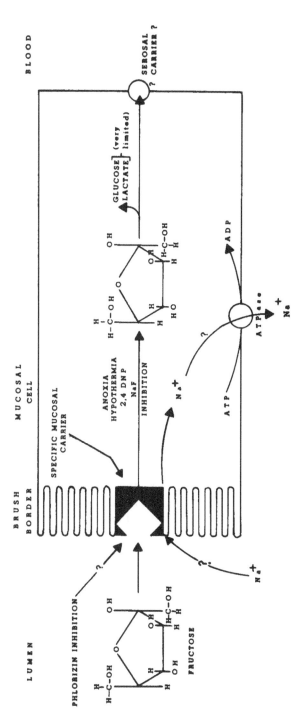

FIGURE 3. Hypothetical model for the transport of fructose by rat small intestine. The transport of fructose is mediated by a carrier located in the brush border that is highly specific for fructose and distinct from the carrier mediating glucose transport. Fructose transport occurs against a concentration gradient and is therefore active, implying the requirement for energy, and thus may be inhibited by antimetabolites such as dinitrophenol (2,4-DNP) and fluoride or by anaerobic conditions or low temperature.[39,40] Active fructose transport may depend to some extent on the maintenance of a Na⁺-gradient as a source of energy. A very small portion of absorbed fructose is converted to glucose or lactate during absorption. Most of the fructose diffuses to the serosal membrane of the epithelial cell where it is absorbed into the blood, perhaps by a carrier-mediated process.

Table 3
ABSORPTION OF
INTRAGASTRICALLY
ADMINISTERED FRUCTOSE
(2.5 g/kg BODY WEIGHT) BY
RATS FED DIETS
CONTAINING 70%
GLUCOSE, SUCROSE, OR
FRUCTOSE FOR 4 WEEKS

Diet	Fructose absorption (μg/mg dry intestine per 2 hr)
Glucose	687 ± 39
Sucrose	867 ± 31[a]
Fructose	842 ± 22[a]

Note: Each value represents the mean ± SEM from seven rats. Statistical significance was obtained by student's *t*-test.

[a]   Significantly greater ($p < 0.01$) than corresponding glucose-fed animals.

Adapted from Vrána, A., Fabry, P., and Kazdová, L., *Physiol. Bohemoslov.*, 26, 225, 1977.

carbohydrate. In rats, the feeding of diets containing either fructose or sucrose appears to produce adaptive increases in the transport system governing fructose absorption. Rats starved for 3 days and then refed for 1 day a diet containing 68% sucrose absorbed 3.3 times more fructose than did similarly starved rats refed a diet containing 85% casein.[10] This increase in fructose absorption was associated with a twofold greater maximal velocity of intestinal sucrase in the sucrose refed rats. A significant twofold increase in the in vivo absorption of fructose in rat intestine was reported following the feeding of a 10% solution of fructose for a period of 3 days.[48] Similarly, rats fed a diet containing 60% fructose as compared to a stock diet for 3 days showed about twice the rate of fructose uptake by intestinal segments.[35] The rate of fructose uptake was not further altered after the rats had received the fructose diet for 15 days. In contrast, the rate of fructose uptake in segments of intestine taken from guinea pigs fed the fructose diet for 3 days was the same as that of animals fed normal laboratory chow.[49] Table 3 presents the comparative levels of the absorption of intragastrically administered fructose (2.5 g/kg body weight) in rats fed diets containing 70% glucose, sucrose, or fructose[50] for 4 weeks. Both sucrose and fructose feeding resulted in about a 25% greater absorption of fructose as compared to glucose feeding ($p < 0.01$). This increase was maximal after feeding the fructose for 2 weeks and remained constant after 8 weeks of feeding. An almost maximal increase of fructose absorption was observed when 25% of the glucose in the glucose-containing diet was replaced by fructose.

The presence of fructose or sucrose in the small intestine of rats not only appears to increase the transport system governing the absorption of fructose but also transport systems mediating the transport of other monosaccharides. A 48-hr intraduodenal perfusion of either 50 m*M* glucose or fructose as compared to an electrolyte solution or 50 m*M* 3-*O*-methylglucose significantly increased the subsequent absorption of 3-*O*-methylglucose.[51] 3-*O*-Methylglucose is a nonmetabolizable glucose analogue that is actively transported by a pathway

Table 4
MONOSACCHARIDE TRANSPORT BY INTESTINAL EPITHELIAL CELLS ISOLATED FROM RATS FED A 54% SUCROSE OR STARCH DIET FOR 8 TO 12 WEEKS

Diet	Monosaccharide transport (mM/5 min)			
	1 mM glucose	1 mM α-methylglucose	1 mM galactose	5 mM fructose
Sucrose	5.6 ± 0.7[a]	12.8 ± 1.7[a]	2.9 ± 0.5	7.5 ± 0.9[a]
Starch	4.1 ± 0.4	10.8 ± 0.9	2.8 ± 0.3	4.9 ± 0.8

Note: Each value represents the mean ± SEM from eight rats.

[a] Significantly greater than corresponding starch value ($p < 0.05$) as determined by a paired difference t-test.

Adapted from Reiser, S., Hallfrisch, J., Putney, J., and Lev, F., *Nutr. Metab.*, 20, 461, 1976.

Table 5
EFFECT OF GLUCOSE- AND FRUCTOSE-CONTAINING DIETS ON THE UPTAKE OF FRUCTOSE AND GLUCOSE BY SEGMENTS OF THE JEJUNUM AND ILEUM

Diet	Uptake (μmol/mℓ tissue H₂O per hour)			
	Fructose		Glucose	
	Jejunum	Ileum	Jejunum	Ileum
30% glucose	34.6 ± 11.9	25.5 ± 9.3	55.2 ± 12.3	36.9 ± 12.7
65% glucose	46.8 ± 9.5[a]	28.9 ± 7.3	71.7 ± 15.4[a]	62.4 ± 17.8[a]
65% fructose	78.1 ± 19.6[a,b]	36.5 ± 6.4[a,b]	65.7 ± 19.0	60.0 ± 28.0[a]

Note: Each value is the mean ± S.D. from six rats.

[a] Significantly greater than 30% glucose diet ($p < 0.05$).
[b] Significantly greater than 65% glucose diet ($p < 0.05$).

Adapted from Bode, C., Eisenhardt, J. M., Haberich, F. J., and Bode, J. C., *Res. Exp. Med.*, 179, 163, 1981.

utilized by glucose.[52] Isolated intestinal epithelial cells from rats fed diets containing 65% sucrose as compared to ground stock diet for 7 to 12 weeks showed a significantly ($p < 0.05$) greater accumulation of 1 mM glucose, 1 mM α-methylglucose, and 5 mM fructose, but not 1 mM galactose.[53] Since it was difficult to attribute the differences in the transport rates found in this study as being due only to differences in the nature of the utilizable carbohydrate present in the two diets, this study was repeated using diets containing either 54% sucrose or starch.[42] Table 4 shows the effect of feeding these different carbohydrates on monosaccharide transport by intestinal epithelial cells after 8 to 12 weeks. In agreement with the previous study,[53] the feeding of sucrose as compared to a starch-based carbohydrate significantly increased the transport of glucose, α-methylglucose, and fructose, but not galactose. In a more recent study,[54] rats were fed for 3 days diets containing 30 or 65% glucose or 65% fructose, and the uptake of fructose and glucose by intestinal segments from the jejunum and ileum was compared (Table 5). The uptake of fructose was significantly greater after the rats consumed the fructose diet than after either of the glucose diets. Glucose

Table 6

THE EFFECT OF PREVIOUS DIET ON FRUCTOSE CONCENTRATION IN
PORTAL AND ARTERIAL BLOOD OF BABOONS BEFORE AND 30 AND 60
MIN AFTER THE INTRAGASTRIC INSTALLATION OF A SUCROSE LOAD
(2 g/kg BODY WEIGHT)

	Blood (mg/100 mℓ)					
	Portal			Arterial		
Diet	0	30	60	0	30	60
Standard (70% starch)	0.2 ± 0.09	30.8 ± 2.27	34.1 ± 5.02	0.1 ± 0.09	11.8 ± 1.30	11.3 ± 1.65
3 weeks, 75% sucrose	0.2 ± 0.09	81.9 ± 8.25	79.4 ± 6.06	0.3 ± 0.10	15.8 ± 2.54	12.8 ± 1.20
6 weeks, 75% sucrose	0.3 ± 0.12	83.3 ± 9.93	82.0 ± 4.18	0.6 ± 0.06	18.7 ± 2.16	15.1 ± 0.83
9 weeks, 75% sucrose	0.1 ± 0.13	78.4 ± 11.58	68.5 ± 2.72	0 ± 0	19.3 ± 3.61	13.9 ± 0.52
3 weeks standard after 75% sucrose	0.5 ± 0.13	50.7 ± 7.77	42.6 ± 2.53	0.3 ± 0.09	13.2 ± 2.15	12.1 ± 1.50

Note: Each value represents the mean ± SEM from six baboons.

Adapted from Crossley, J. N. and Macdonald, I., *Nutr. Metab.*, 12, 171, 1970.

uptake was increased by about the same amount in rats fed either the 65% glucose or fructose diets as compared to the 30% glucose diet.

The feeding of sucrose also appears to increase adaptively the absorption of fructose in primates. Baboons with chronic portal venous cannuli were fed a standard diet containing 70% starch and then fed a diet containing 75% sucrose for 9 weeks.[55] The levels of fructose and glucose appearing in the portal venous and femoral arterial blood after a sucrose test meal were used as an indicator of monosaccharide absorption. Table 6 presents the blood fructose levels before and 30 and 60 min following the sucrose load while the baboons were fed the standard and high-sucrose diets. The high-sucrose diet produced a significant increase in blood fructose concentration as compared to the standard diet. In the portal blood the fructose concentration appeared to be increased after the sucrose load even 3 weeks after returning to the standard diet. Since the high-sucrose diet did not have any significant effect on the levels of portal or arterial blood glucose following the sucrose meal, these results were interpreted as being due to an increased absorption of fructose rather than an increased digestion of sucrose following adaptation to the high-sucrose diet.

If the results obtained with rats and baboons are relevant to humans, it appears that adaptation to diets high in sucrose or fructose increases the absorptive efficiency of fructose.

The mechanism by which dietary components such as sucrose or fructose can produce adaptive increases in monosaccharide transport is not apparent. Changes in the rate of absorption may include changes in the carrier density at either the brush border or basolateral membranes, efficiency of these carriers, unstirred layer thickness, energy required for active transport, and number of transporting cells. Since transport carriers are mainly conceptual and have not been isolated per se, their number and efficiency are determined by kinetic approaches. Changes in the V_{max} are considered to represent changes in the carrier density and changes in the K_m represent changes in the efficiency with which the carrier binds the transported substance.

As the transport of fructose and glucose appears to be mediated by distinct intestinal carriers, it seems unlikely that the adaptive increases in fructose and glucose transport produced by the feeding of sucrose or fructose are due to changes in carrier density or efficiency. A kinetic evaluation of the effect of the feeding of glucose, galactose, and α-methylglucose on the absorption of these monosaccharides by rat intestine revealed complex interactions that were not compatible with the assumption that dietary adaptation was affected through changes in the number or efficiency of a single transport pathway.[56] One possible mechanism by which the feeding of sucrose or fructose may adaptively increase both fructose and glucose absorption is based on a relationship between sucrase activity and the transport of its component monosaccharides. As shown above, the feeding of diets high in sucrose produces adaptive increases in sucrase activity. The addition of the sucrase-isomaltase complex to lipid bilayers has been found to increase greatly the permeability of sucrose through the membrane.[57] It also has been shown that the kinetic properties of the activation of sucrase by sodium ions closely resemble the kinetic properties of the activation of monosaccharide transport by sodium ions.[58] These findings suggest a close association between sucrase activity and monosaccharide transport in the intestinal brush border and raise the possibility that factors which influence the activity of one of these processes may also affect the other.

It is known that the intake of diets high in sucrose or fructose increases the activities of the fructose-metabolizing enzymes fructokinase and fructose-1-phosphate aldolase in the intestine of rats[59-61] and humans.[62] It is therefore possible that the increased absorption of fructose observed after fructose feeding may be due to a favorable metabolic gradient produced by the increased metabolism of fructose in the enterocyte. As a consequence of this increased metabolism of fructose, ATP levels may be increased, providing energy not only for the absorption of fructose but also for the absorption of glucose. However, it has been reported that fructose feeding produces an increase in fructose transport in rat small intestine when fructose metabolism is unaffected.[35]

REFERENCES

1. U.S. Department of Agriculture, Sugar and Sweetener Outlook and Situation Report No. SSRV10N1, Economic Research Service, U.S. Government Printing Office, Washington, D.C., March 31, 1985.
2. **Johnson, C. F.,** Disaccharidase: location in hamster intestinal brush border, *Science,* 155, 1670, 1967.
3. **Johnson, C. F.,** Hamster intestinal brush-border surface particles and their function, *Fed. Proc., Fed. Am. Soc. Exp. Biol.,* 28, 26, 1969.
4. **Kolínská, J. and Semenza, G.,** Studies on isolated sucrase and on isomaltase sugar transport. V. Isolation and properties of sucrase-isomaltase from rabbit small intestine, *Biochim. Biophys. Acta,* 146, 181, 1967.
5. **Cummins, D. L., Gitzelmann, R., Lindenmann, J., and Semenza, G.,** Immunochemical study of isolated human and rabbit intestinal sucrase, *Biochim. Biophys. Acta,* 160, 396, 1968.
6. **Hauri, H.-P., Roth, J., Sterchi, E. E., and Lentze, M. J.,** Transport to cell surface of intestinal sucrase-isomaltase is blocked in the Golgi apparatus in a patient with congenital sucrase-isomaltase deficiency, *Proc. Natl. Acad. Sci. U.S.A.,* 82, 4423, 1985.
7. **Riby, J. E. and Kretchmer, N.,** Effect of dietary sucrose on synthesis and degradation of intestinal sucrase, *Am. J. Physiol.,* 246, G757, 1984.
8. **Hauri, H.-P., Quaroni, A., and Isselbacher, K. J.,** Biogenesis of intestinal plasma membrane: post translational route and cleavage of sucrase-isomaltase, *Proc. Natl. Acad. Sci. U.S.A.,* 76, 5183, 1979.
9. **Blair, D. G. R., Yakimets, W., and Tuba, J.,** Rat intestinal sucrase. II. The effects of rat age and sex and of diet on sucrase activity, *Can. J. Biochem. Physiol.,* 41, 917, 1963.
10. **Daren, J. J., Broitman, S. A., and Zamcheck, N.,** Effect of diet upon intestinal disaccharidases and disaccharide absorption, *J. Clin. Invest.,* 46, 186, 1967.
11. **Reddy, B. S., Pleasants, J. R., and Wostman, B. S.,** Effect of dietary carbohydrate on intestinal disaccharidases in germfree and conventional rats, *J. Nutr.,* 95, 413, 1968.
12. **Ulshen, M. H. and Grand, R. J.,** Site of substrate stimulation of jejunal sucrase in the rat, *J. Clin. Invest.,* 64, 1097, 1979.
13. **Kimura, T., Matsumoto, Y., and Yoshida, A.,** Dietary sucrose-mediated changes in jejunal sucrase activity of rats, *J. Nutr. Sci. Vitaminol.,* 26, 585, 1980.
14. **Raul, F., Simon, P. M., Kedinger, M., Grenier, J. F., and Haffen, K.,** Effect of sucrose refeeding on disaccharidase and aminopeptidase activities of intestinal villus and crypt cells in adult rats. Evidence for a sucrose-dependent induction of sucrase in crypt cells, *Biochim. Biophys. Acta,* 630, 1, 1980.

15. **Yamada, K., Bustamante, S., and Koldovsky, O.,** Dietary-induced rapid increases of rat jejunal sucrase and lactase activity in all regions of the villus, *FEBS Lett.,* 129, 89, 1981.

16. **Bustamante, S., Gasparo, M., Kendall, K., Coates, P., Brown, S., Somawane, B., and Koldovsky, O.,** Increased activity of rat intestinal lactase due to increased intake of α-saccharides (starch, sucrose) in isocaloric diets, *J. Nutr.,* 111, 943, 1981.

17. **Koldovsky, O., Bustamante, S., and Yamada, K.,** Adaptability of lactase and sucrase activity in jejunoileum of adult rats to changes in intake of starch, sucrose, lactose, glucose, fructose and galactose, in *Mechanisms of Intestinal Adaptation,* Robinson, J. W. L., Dowling, R. H., and Riecken, E.-O., Eds., MTP Press, Hingham, England, 1982, chap. 12.

18. **Chung-Ja, M. C. and Randall, H. T.,** Effects of substitution of glucose-oligosaccharides by sucrose in a defined formula diet on intestinal disaccharidases, hepatic lipogenic enzymes and carbohydrate metabolism in young rats, *Metabolism,* 31, 57, 1982.

19. **Gorostiza, E., Marche, C., Broyart, J. P., Balmain, N., and Cezard, J. P.,** Influence of starvation on sucrase regulation by dietary sucrose in the rat, *Am. J. Clin. Nutr.,* 40, 1017, 1984.

20. **Rosensweig, N. S. and Herman, R. H.,** Control of jejunal sucrase and maltase activity by dietary sucrose or fructose in man, *J. Clin. Invest.,* 47, 2253, 1968.

21. **Kolínská, J. and Kraml, J.,** Separation and characterization of sucrase-isomaltase complex and of glucoamylase of rat intestine, *Biochim. Biophys. Acta,* 284, 235, 1972.

22. **Rosensweig, N. S. and Herman, R. H.,** Time response of jejunal sucrase and maltase activity to a high sucrose diet in normal man, *Gastroenterology,* 56, 500, 1969.

23. **James, W. P. T., Alpers, D. H., Gerber, J. E., and Isselbacher, K. J.,** The turnover of disaccharidases and brush border proteins in rat intestine, *Biochim. Biophys. Acta,* 230, 194, 1971.

24. **Yamada, K., Goda, T., Bustamante, S., and Koldovsky, O.,** Different effect of starvation on activity of sucrase and lactase in rat jejunoileum, *Am. J. Physiol.,* 244, G449, 1983.

25. **Goda, T., Yamada, K., Bustamante, S., and Koldovsky, O.,** Dietary-induced rapid decrease in microvillar carbohydrase activity in rat jejunoileum and degradation of intestinal sucrase, *Am. J. Physiol.,* 246, G757, 1984.

26. **Cezard, J. P., Broyard, J. P., Cussinier-Gleizes, P., and Matthien, H.,** Sucrase-isomaltase regulation by dietary sucrose in the rat, *Gastroenterology,* 84, 18, 1983.

27. **Raul, F. and van der Decken, A.,** Modulation of RNA polymerase activities in the intestine of adult rats by dietary sucrose, *J. Nutr.,* 113, 2134, 1983.

28. **Simon, P. M., Kedinger, M., Raul, F., Grenier, J. F., and Hasfen, K.,** Organ culture of suckling rat intestine. Comparative study of various hormones on brush border enzymes, *In Vitro,* 18, 339, 1981.

29. **Menard, D., Malo, C., and Calvert, R.,** Insulin accelerates the development of intestinal brush border hydrolase activities of suckling mice, *Dev. Biol. Med.,* 85, 150, 1984.

30. **Lee, P. C., Brooks, S., and Lebenthal, E.,** Effect of glucose and insulin on small intestinal brush border enzymes in fasted rats, *Proc. Soc. Exp. Biol. Med.,* 173, 372, 1983.

31. **Vrána, A. and Fábry, P.,** Metabolic effects of high sucrose or fructose intake, *World Rev. Nutr. Diet.,* 42, 56, 1983.

32. **Cori, C. F.,** The fate of sugars in the animal body. I. The rate of absorption of hexoses and pentoses from the intestinal tract, *J. Biol. Chem.,* 66, 691, 1925.

33. **Kiyasu, J. T. and Chaikoff, I. L.,** On the manner of transport of absorbed fructose, *J. Biol. Chem.,* 224, 935, 1957.

34. **Ginsburg, V. and Hers, H.-G.,** On the conversion of fructose to glucose by guinea pig intestine, *Biochim. Biophys. Acta,* 38, 427, 1960.

35. **Mavrias, D. A. and Mayer, R. J.,** Metabolism of fructose in the small intestine. I. The effect of fructose feeding on fructose transport and metabolism in rat small intestine, *Biochim. Biophys. Acta,* 291, 531, 1973.

36. **Leveille, G. A., Akinbami, K., and Ikediobi, C. O.,** Fructose absorption and metabolism by the growing chick, *Proc. Soc. Exp. Biol. Med.,* 135, 483, 1970.

37. **Holdsworth, C. D. and Dawson, A. M.,** Absorption of fructose in man, *Proc. Soc. Exp. Biol. Med.,* 118, 142, 1965.

38. **Cook, G. C.,** Absorption and metabolism of D(-)fructose in man, *Am. J. Clin. Nutr.,* 24, 1302, 1971.

39. **Gracey, M., Burke, V., and Oshin, A.,** Active intestinal transport of D-fructose, *Biochim. Biophys. Acta,* 266, 397, 1972.

40. **Macrae, A. R. and Neudoerffer, T. S.,** Support for the existence of an active transport mechanism of fructose in the rat, *Biochim. Biophys. Acta,* 288, 137, 1972.

41. **Michaelis, O. E., IV, Hallfrisch, J., Putney, J., and Reiser, S.,** Intestinal uptake by starved-refed rats of sugars derived from disaccharides and from their monosaccharide equivalents, *Nutr. Rep. Int.,* 13, 107, 1976.

42. **Reiser, S., Hallfrisch, J., Putney, J., and Lev, F.,** Enhancement of intestinal sugar transport by rats fed sucrose as compared to starch, *Nutr. Metab.,* 20, 461, 1976.

43. **Schultz, S. G. and Strecker, C. K.**, Fructose influx across the brush border of rabbit ileum, *Biochim. Biophys. Acta*, 211, 586, 1970.
44. **Guy, M. J. and Deren, J. J.**, Selective permeability of the small intestine for fructose, *Am. J. Physiol.*, 221, 1051, 1971.
45. **Honegger, P. and Semenza, G.**, Multiplicity of carriers for free glucalogues in hamster small intestine, *Biochim. Biophys. Acta*, 318, 390, 1973.
46. **Sigrist-Nelson, K. and Hopfer, U.**, A distinct D-fructose transport system in isolated brush border membrane, *Biochim. Biophys. Acta*, 367, 247, 1974.
47. **Reiser, S. and Hallfrisch, J.**, Stimulation of neutral amino acid transport by fructose in epithelial cells isolated from rat intestine, *J. Nutr.*, 107, 767, 1977.
48. **Crouzoulon-Bourcart, C., Crouzoulon, G., and Pérès, G.**, Recherches sur l'absorption intestinale des hexoses. II. Effets de quelques régimes simples sur l'absorption intestinale du fructose in vivo, *C. R. Soc. Biol.*, 165, 1071, 1972.
49. **Mavrias, D. A., and Meyer, J. B.**, Metabolism of fructose in the small intestine. II. The effect of fructose feeding on fructose transport and metabolism in guinea pig small intestine, *Biochim. Biophys. Acta*, 291, 538, 1973.
50. **Vrána, A., Fábry, P., and Kazdová, L.**, Diet-induced adaptation of intestinal fructose absorption in the rat, *Physiol. Bohemoslov.*, 26, 225, 1977.
51. **Roy, C. C. and Dubois, R.**, Monosaccharide induction of 3-O-methylglucose transport through the rat jejunum, *Proc. Soc. Exp. Biol. Med.*, 139, 883, 1972.
52. **Wilson, T. H. and Landau, B. R.**, Specificity of sugar transport by the intestine of the hamster, *Am. J. Physiol.*, 198, 99, 1960.
53. **Reiser, S., Michaelis, O. E., IV, Putney, J., and Hallfrisch, J.**, Effect of sucrose feeding on the intestinal transport of sugars in two strains of rats, *J. Nutr.*, 105, 894, 1975.
54. **Bode, C., Eisenhardt, J. M., Haberich, F. J., and Bode, J. C.**, Influence of feeding fructose on fructose and glucose absorption in rat jejunum and ileum, *Res. Exp. Med.*, 179, 163, 1981.
55. **Crossley, J. N. and Macdonald, I.**, The influence in male baboons of a high sucrose diet on portal and arterial levels of glucose and fructose following a sucrose meal, *Nutr. Metab.*, 12, 171, 1970.
56. **Debnam, E. S. and Levin, R. J.**, Influence of specific dietary sugars on the jejunal mechanisms for glucose, galactose, and α-methylglucoside absorption: evidence for multiple sugar carriers, *Gut*, 17, 92, 1976.
57. **Storelli, C., Vogeli, H., and Semenza, G.**, Reconstitution of a sucrase-mediated sugar transport system in lipid membrane, *FEBS Lett.*, 24, 287, 1972.
58. **Schultz, S. G. and Curran, P. F.**, Coupled transport of sodium and organic solutes, *Physiol. Rev.*, 50, 637, 1970.
59. **Stifel, F. B., Rosensweig, N. S., Zakim, D., and Herman, R. H.**, Dietary regulation of glycolytic enzymes. I. Adaptive changes in rat jejunum, *Biochim. Biophys. Acta*, 170, 221, 1968.
60. **Crouzoulon, G. and Dandrifosse, G.**, Effect of a high fructose diet on the activities of enzymes implicated in the conversion of fructose to glucose by the liver and the intestinal mucosa of the rat, *Biochem. Syst. Ecol.*, 3, 273, 1975.
61. **Bode, C., Bode, J. C., Ohta, W., and Martini, G. A.**, Adaptive changes of the activity of enzymes involved in fructose metabolism in the liver and jejunal mucosa of rats following fructose feeding, *Res. Exp. Med.*, 178, 55, 1980.
62. **Rosensweig, N. S., Stifel, F. B., Herman, R. H., and Zakim, D.**, The dietary regulation of glycolytic enzymes. II. Adaptive changes in human jejunum, *Biochim. Biophys. Acta*, 170, 228, 1968.

Chapter 3

METABOLISM

I. INTRODUCTION

In man and rat fructose is metabolized primarily in the liver,[1,2] but the intestinal mucosa[3] and kidney[4] also contain enzymes required for the specific metabolism of fructose. Utilization of fructose in peripheral extrahepatic tissues is minimal.[5,6] Glucose in the blood inhibits the phosphorylation of fructose by hexokinase.[7,8] Hexokinase can phosphorylate a number of hexoses including glucose, fructose, and mannose, but in the brain has 20 times greater affinity for glucose than for fructose.[9] Glucokinase, which is present only in the liver, is specific for glucose.[10] Fructokinase, which is specific for fructose, is also found in the liver.

In the small intestine fructose transport is slower than glucose transport.[11] Fructose transport in the intestine of the rat is carrier-mediated.[12] In human intestinal mucosa there is a limited amount of conversion of fructose to lactic acid and glucose during absorption.[13] Bode et al.[14] studied the absorption of both fructose and glucose in the jejunum and ileum of the rat. They exposed a defined area of mucosa to a medium containing ^{14}C-labeled glucose or fructose and then determined the amount of uptake of the two sugars. Uptake of both fructose and glucose was greater in the jejunum than the ileum and was influenced by previous diet of the rats (Table 1). When rats were fed 65% glucose, glucose uptake of the jejunum was significantly greater than when rats consumed a diet containing 30% glucose. Consuming a diet of 65% fructose stimulated fructose uptake in both the jejunum and the ileum. In the rat intestine, since there is no glucose-6-phosphatase, there is little conversion of fructose to glucose.

II. INITIAL STEPS OF FRUCTOSE METABOLISM

The initial step in the metabolism of fructose is shown in Figure 1. Fructose is phosphorylated by fructokinase[16-19] to fructose-1-phosphate. This reaction requires Mg^{++}, ATP, and K^+.[20,21] Fructokinase can also catalyze the phosphorylation of some other ketoses, including galactoheptulose, sorbose, tagatose, and xylulose. Affinity is dependent on the amount of potassium.[19,20] Fructose-1-phosphate accumulates rapidly in the liver as a result of the phosphorylation of fructose by fructokinase.[21]

The second reaction in the metabolism of fructose is the splitting of fructose-1-phosphate by fructose-1-phosphate aldolase to dihydroxyacetone phosphate and glyceraldehyde (Figure 2).[22,23] This enzyme can also cleave fructose-1,6-diphosphate and work in the opposite direction to condense the two triose phosphates, dihydroxyacetone phosphate and glyceraldehyde-3-phosphate, into fructose-1,6-diphosphate.[24] Aldolases occurring in brain and muscle have a much lower affinity for fructose-1-phosphate than does liver aldolase,[25] in which the affinity of the enzyme for fructose-1-phosphate is not different from its affinity for fructose-1,6-diphosphate.[26] Formation of dihydroxyacetone phosphate and glyceraldehyde from fructose-1-phosphate was confirmed by labeling fructose with ^{14}C in both the 1-C and 6-C positions.[27]

Figure 3 depicts the diverse fates of dihydroxyacetone phosphate. It can be isomerized to glyceraldehyde phosphate and continue through the glycolytic pathway[28,29] to yield ultimately either lactic acid (anaerobic conditions) or acetyl CoA (aerobic conditions) from pyruvate. The acetyl CoA could then enter the TCA cycle or be used for fatty acid synthesis. Dihydroxyacetone phosphate can also be reduced to glycerol-3-phosphate which can provide the glycerol moiety of triglycerides. Alternatively, dihydroxyacetone phosphate can be con-

Table 1

UPTAKE OF D-FRUCTOSE AND D-GLUCOSE IN JEJUNAL AND ILEAL SEGMENTS FROM RATS FED GLUCOSE- OR FRUCTOSE-CONTAINING DIETS

	Uptake (μmol/mℓ tissue H_2O per hour)			
	D-Fructose		D-Glucose	
Diet	Jejunum	Ileum	Jejunum	Ileum
30% glucose	34.6[a]	25.5[a]	55.2[a,d]	36.9[a]
65% glucose	46.8[b,d]	28.9[a]	71.7[b]	62.4[b]
65% fructose	78.1[c,d]	36.5[b]	65.7[b]	60.0[b]

[a-c] Values within a column are significantly different if they do not share a common superscript ($p < 0.05$).

[d] Uptake in jejunum significantly greater than uptake in ileum of the same group ($p < 0.05$).

Adapted from Bode, C., Eisenhardt, J. M., Haberich, F. J., and Bode, J. C., *Res. Exp. Med.*, 179, 163, 1981.

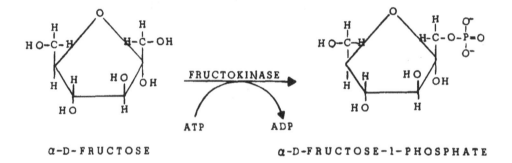

FIGURE 1.   Phosphorylation of fructose.

densed with glyceraldehyde-3-phosphate by fructose-1,6-diphosphate aldolase to form fructose-1,6-diphosphate and ultimately glucose or glycogen. Förster[30] has measured glycogen storage in rats infused with glucose, fructose, xylitol, and sorbitol. About 50% of the infused glucose was stored as glycogen. Glycogen deposition was greater in animals infused with fructose and the two sugar alcohols than in those infused with glucose.

III. THREE FATES OF GLYCERALDEHYDE

There are three possible fates of the glyceraldehyde formed from fructose-1-phosphate (Figure 4).

1.   D-Glyceraldehyde can be phosphorylated to glyceraldehyde-3-phosphate by triokinase.[23] This enzyme has been purified from a number of sources including livers of rats,[31] cattle,[32] and guinea pigs.[33] This enzyme appears to be specific to the metabolism of fructose. Triokinase has been reported to increase significantly in animals fed a fructose-containing diet.[34,35] Phosphorylated glyceraldehyde can continue through the glycolytic pathway to form pyruvate or be condensed with dihydroxyacetone phosphate to form fructose-1,6-diphosphate and ultimately converted to glucose or stored as

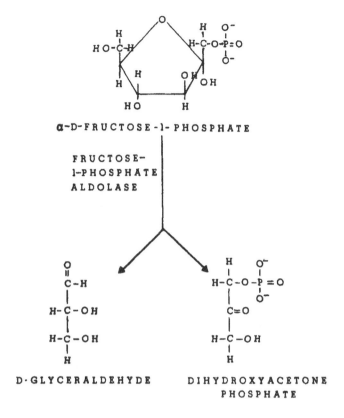

FIGURE 2.   Splitting of fructose-1-phosphate.

glycogen. In the liver, glyceraldehyde-3-phosphate is the major product formed from glyceraldehyde.[11]

2.   Glyceraldehyde can be converted to glycerate and then enter glycolysis after phosphorylation to 2-phosphoglycerate. A number of researchers[21,22,36-39] have proposed this fate on the basis of the accumulation of glycerate in the liver after fructose infusion.[40] There has been considerable evidence to support the contention that triokinase is the enzyme which phosphorylates the glycerate.[34,35,41]

3.   The third fate of glyceraldehyde formed from fructose is reduction to glycerol by NADH-dependent alcohol dehydrogenase[42,43] or by NADPH-dependent aldose reductase.[42,44] The glycerol could then be phosphorylated[45] to glycerol-3-phosphate and either converted to dihydroxyacetone phosphate or to the glycerol moiety of triglycerides, phospholipids, and other lipids. In the kidney, conversion to glycerol may be the primary fate of glyceraldehyde formed from fructose.[11]

IV. IMPORTANCE OF FRUCTOSE-1-PHOSPHATE

One of the major consequences of infusions of fructose is the accumulation of high levels of fructose-1-phosphate.[46-49] It is the accumulation of this compound which results in the adverse affects in hereditary fructose intolerance (see Chapter 4). The accumulation of this phosphorylated compound causes sequestering of phosphorus and depletion of intracellular concentrations of phosphorus and ultimately ATP.[50-54] This depletion has been reported in rats[50,51] as well as man.[52] The reduction in ATP affects many enzymatic processes, decreasing free energy[53] and inhibiting protein synthesis[50] and growth.[54]

Fructose infusions were reported to decrease both ATP and inorganic phosphate levels to about one fourth to one third the normal levels.[50] The depletion of inorganic phosphorus is

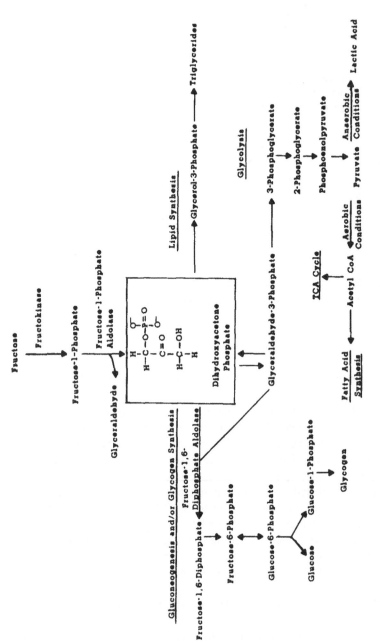

FIGURE 3.   Fates of dihydroxyacetone phosphate.

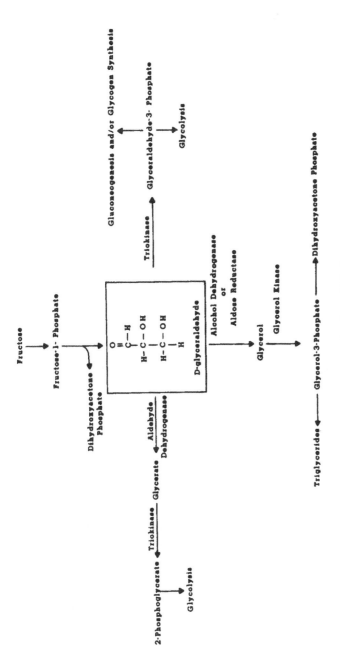

FIGURE 4.   Fates of D-glyceraldehyde.

Table 2
PLASMA INORGANIC
PHOSPHORUS RESPONSES OF
HYPERINSULINEMIC AND
NORMAL MEN CONSUMING
THREE LEVELS OF FRUCTOSE

Diet (% fructose)	Response (mg/dℓ) at various times (hr) after ingestion				
	0	**1/2**	**1**	**2**	**3**
Control					
0	3.7	3.2	3.0	2.7	3.1
7.5	3.6	3.2	2.9	2.7	3.0
15.0	3.7	3.3	2.8	2.3	2.9
Hyperinsulinemic					
0	3.8	3.3	2.7	2.2	2.8
7.5	4.2	3.5	3.6	3.3	3.3
15.0	4.1	4.3	4.2	4.2	3.6

Adapted from Hallfrisch, J., Ellwood, K., Michaelis, O. E., IV, Reiser, S., and Prather, E. S., *J. Am. Coll. Nutr.*, 5, 61, 1986.

apparently the factor which determines the amount of ATP depletion, not the increase of fructose-1-phosphate.[55] Hypophosphatemia occurs in a number of clinical situations, including alcoholism. Infusion of fructose in these situations can have dire results.[56] Acute phosphate depletion can cause renal or heart failure and disturbances in leukocyte and thrombocyte functions, and chronic phosphate depletion can cause skeletal abnormalities. While fructose infusions have caused significant depletion of phosphate and ultrastructural changes in the liver,[57,58] some of these changes were also reported after glucose infusions. Low levels of ATP per se do not appear to produce changes in liver ultrastructure, since there was no visible decrease in viability of liver parenchymal cells after maintaining ATP levels at 20 to 25% of normal for 36 to 48 hr.[59] Inflammation of the liver characterized by jaundice is often due to toxic agents. Increases in blood bilirubin have been observed when xylitol was infused into humans.[30] It was therefore of interest to determine whether sugars such as fructose and other polyols could produce changes characteristic of liver damage. Rapid infusion of 1.5 g/kg body weight of fructose, glucose, xylitol, or sorbitol produced small and identical increases in serum bilirubin of healthy subjects.[30,60]

Hallfrisch et al.[61] found phosphorus responses after a sucrose load in normal men to be unaffected by the level of fructose in the diet, but in hyperinsulinemic men the phosphorus levels were higher than in normal men. Diets containing 0, 7.5, or 15% of the calories as fructose for 5 weeks each were consumed by 12 hyperinsulinemic and 12 normal men. At the end of each time period, 2 g sucrose per kilogram body weight was given in an oral solution. Plasma inorganic phosphorus was measured at fasting and $^1/_2$, 1, 2, and 3 hr after sucrose ingestion (Table 2). In normal men phosphorus levels decreased from fasting to 2 hr after sucrose consumption and then began to increase. Normal men who consumed the highest level of fructose experienced the greatest drop in phosphorus (3.7 to 2.3 mg/dℓ), indicating that more phosphorus was being sequestered. However, when the hyperinsulinemic men consumed 7.5 and 15.0% fructose, fasting levels of phosphorus were higher and decreased more slowly after the sucrose load. This effect was also reported after adaptation to the fructose-containing disaccharide sucrose in hyperinsulinemic,[62] but not normal[63] men

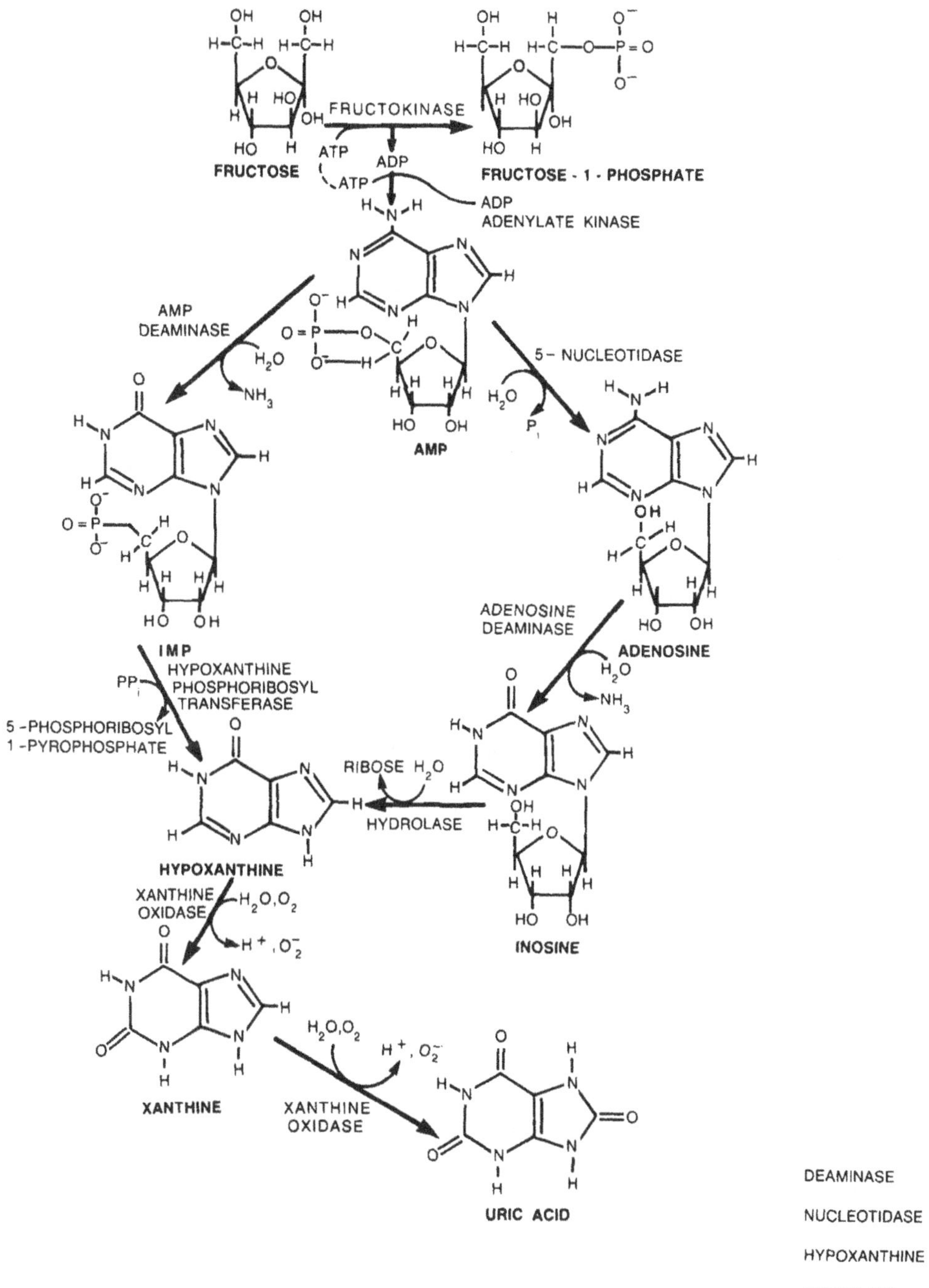

FIGURE 5.   Nucleotide metabolism resulting from fructose.

and women. This delay in the depletion of phosphorus after adaptation to fructose-containing diets indicates an abnormality of metabolism in these hyperinsulinemic subjects.

Decreased availability of ATP and phosphorus also results in the accumulation of AMP and subsequent activation of AMP deaminase. This enzyme is responsible for the conversion of AMP to IMP, which is ultimately converted to uric acid (Figure 5). It is by this mechanism that fructose infusion or feeding can result in increased levels of uric acid.[64] Hyperuricemia, although not necessarily a direct cause, is associated with increased risk of heart disease[65]

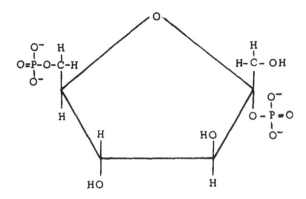

FIGURE 6.   Structure of α-ᴅ-fructose-2,6-biphosphate.

and diabetes.[66] Both infusion[30] and oral administration of fructose increased uric acid.[67,70] Fructose infusions also increased uric acid levels of nonketogenic and ketogenic diabetics[71] and hypertensives.[72] Studies in which fructose was fed as a part of the regular diet have produced variable results. Uric acid was increased in both normal and especially hereditary fructose-intolerant children fed 0.33 to 1.3 g/kg body weight fructose for 3 to 5 days.[73] The Turku sugar studies reported no difference in fasting or casual uric acid levels before or after feeding fructose, sucrose, or xylitol for 22 months.[74] However, all three sugars would be expected to increase uric acid levels. Since initial levels were not measured in the fasting state, it is difficult to determine whether changes in uric acid levels occurred. Turner et al.[75] fed 20% of the carbohydrate calories as fructose or dextromaltose in both fat-free and fat-containing formula diets for 2 weeks. Serum uric acid levels were not affected. Consumption of only 7.5 or 15.0% of total calories as fructose for 5 weeks resulted in significant increases in uric acid responses of both normal and hyperinsulinemic men, but fasting levels were only slightly increased.[61] Uric acid concentrations of hyperinsulinemic men were higher than those of controls. Consumption of fructose-containing sucrose also has increased uric acid levels in normal,[63] hyperinsulinemic,[62] obese, and gouty[76] people. In the obese and controls, reduced renal clearance may contribute to the hyperuricemia, but in gouty patients sucrose increases water production.[75]

V. ROLE OF FRUCTOSE-2,6-BIPHOSPHATE

A recently discovered form of fructose (fructose-2,6-biphosphate, Figure 6) may serve as an important regulator of carbohydrate metabolism in the liver. The role of this compound has recently been reviewed[77] and will be discussed briefly in this chapter. Van Schaftingen and Hers[78] were the first to tentatively identify this labile compound. The synthesis and degradation of fructose-2,6-biphosphate are catalyzed by a single enzyme complex (6-phosphofructo-2-kinase/fructose-2,6-biphosphatase).[79-82] This enzyme is probably regulated by cAMP-dependent phosphorylation and dephosphorylation. The dephosphorylation has not been characterized.[83] The level of fructose-2,6-biphosphate appears to affect the activities of two important regulatory enzymes involving carbohydrate metabolism in the liver. A summary of the reactions by which fructose-2,6-biphosphate can regulate glycolysis and gluconeogenesis is shown in Figure 7. The amount of fructose-2,6-biphosphate is regulated by the enzyme complex 6-phosphofructo-2-kinase/fructose-2,6-biphosphatase. cAMP stimulates the fructose-2,6-biphosphatase which then forms fructose-6-phosphate and reduces the levels of fructose-2,6-biphosphate. The reduction in the levels of fructose-2,6-biphosphate stimulates fructose-1,6-diphosphatase, which favors the conversion of fructose-1,6-diphosphate to fructose-6-phosphate, an intermediate of gluconeogenesis. In contrast, an increase

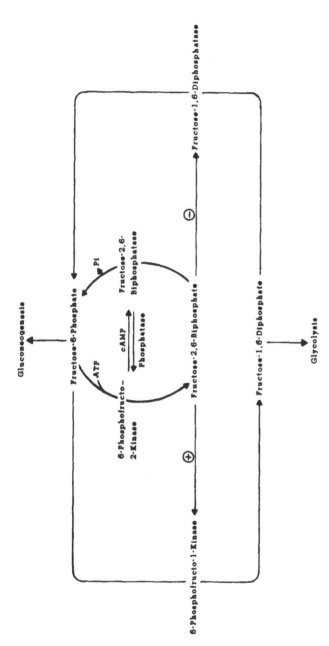

FIGURE 7.   Regulation of glycolysis and gluconeogenesis by fructose-2,6-biphosphate. (Adapted from Claus, T. H., El-Maghrabi, M. R., Regen, D. M., Stewart, H. B., McGrane, M., Kountz, P. D., Nyfeler, F., Pilkis, J., and Pilkis, S. J., *Curr. Top. Cell. Regul.*, 23, 57, 1984.)

in fructose-2,6-biphosphate stimulates 6-phosphofructo-1-kinase, which converts fructose-6-phosphate to fructose-1,6-diphosphate, the committed step in glycolysis.

Hue[84] studied the role of fructose-2,6-biphosphate in the anaerobic stimulation of glycolysis in isolated hepatocytes from starved rats. Hepatocytes were incubated with increasing levels of glucose to stimulate lactate production. Fructose-2,6-biphosphate was reduced during anaerobic glycolysis and therefore is not responsible for the stimulation of glycolysis by anoxia. Neely et al.[85] also measured the level of fructose-2,6-biphosphate in livers of starved rats as well as rats made diabetic with alloxan, in order to determine the role of insulin on the regulation of this metabolite. Fructose-2,6-biphosphate levels were 10% of normal in livers of both starved and diabetic animals. Insulin administration restored normal levels in diabetic rats. Refeeding carbohydrate to starved rats increased levels 2.5 times higher than in livers of fed rats. These changes in levels of fructose-2,6-biphosphate parallel rates of glycolysis and gluconeogenesis and provide further evidence that this metabolite regulates carbohydrate metabolism in the liver.

The relationships between insulin, glucagon, cAMP, and epinephrine in metabolism of fructose-2,6-biphosphate were examined in hepatocytes from both diabetic and normal rats by Pilkis et al.[80] Glucagon and epinephrine lower levels of fructose-2,6-biphosphate in part by the action of cAMP on a protein kinase which phosphorylates and thereby activates the enzyme responsible for degradation of fructose-2,6-biphosphate. Insulin inhibits this action. Insulin decreased fructose-2,6-biphosphatase and increased 6-phosphofructo-2-kinase activities in treated hepatocyte extracts. Diabetic and starved rats had reduced levels of hepatic fructose-2,6-biphosphate. Adding glucose to the medium increased the levels of this metabolite in cells from starved and fed rats, but not diabetic rats, indicating the role of insulin in metabolism of fructose-2,6-biphosphate.

Bartrons et al.[86] also studied the hormonal control of fructose-2,6-biphosphate in isolated rat hepatocytes. They found no difference in the sensitivity of hepatocytes to glucagon in the interconversion between phosphorylase a and b, the active and inactive forms of pyruvate kinase, and 6-phosphofructo-2-kinase and fructose-2,6-biphosphatase. However, their results contrast with those of El-Maghrabi et al.,[81] who found ten times greater sensitivity for interconversion of 6-phosphofructo-2-kinase and fructose-2,6-biphosphatase than between pyruvate kinase and the phosphorylases. Bartrons et al.[86] concluded that the hormonally induced reduction of fructose-2,6-biphosphate was due to cAMP.

El-Maghrabi et al.[87] characterized the cAMP-dependent protein kinase from rat liver by partially purifying 6-phosphofructo-2-kinase by polyethylene glycol precipitation, DEAE-cellulose chromatography, $(NH_4)_2SO_4$ fractionation, gel filtration, and Sephadex chromatography. The purified enzyme was then incubated with the catalytic subunit of the protein kinase and radioactive ATP. Phosphorus was incorporated into a protein subunit with a molecular weight of 49,000, which was associated with an inhibition of 6-phosphofructo-2-kinase activity and decreased affinity for fructose-6-phosphate.

Reinhart[88] suggested that fructose-2,6-biphosphate influences the aggregation of 6-phosphofructo-1-kinase in rat liver, and in this way affects the hepatic metabolism of carbohydrate. The smallest active form of the enzyme is a tetramer with a molecular weight of about 82,000, but very large aggregates with molecular weights in the millions can form. The large aggregates appear to be more active than the tetramers. By using pyrenebutyric acid, a fluorescent probe bound to 6-phosphofructo-1-kinase, fructose-2,6-biphosphate was found to slow the dissociation of the enzyme and promote reassociation. Increasing the activity of 6-phosphofructo-1-kinase would then promote glycolysis.

The bifunctional enzyme which catalyzes the formation and degradation of fructose-2,6-biphosphate was isolated and its properties further characterized by Pilkis et al.[89] Fructose-2,6-biphosphate was incubated with purified 6-phosphofructo-2-kinase/fructose-2,6-biphosphatase isolated from rat liver in the presence of fructose-6-phosphate, ATP, and inorganic phosphorus. A two-site catalyzation model of the action of the enzyme has been proposed (Figure 8). This model accounts for the formation of fructose-2,6-biphosphate from fructose-6-phosphate, and for phosphorylation and dephosphorylation of the enzyme.

Although most of the research examining the influence of fructose-2,6-biphosphate on carbohydrate metabolism has been done with liver cells from normal and diabetic rats, Hue et al.[90] have also studied hepatocytes from obese rats. Genetically obese (fa/fa) rats had higher levels of fructose-2,6-biphosphate, fructose-6-phosphate, glycogen, and pyruvate kinase and phosphofructokinase activities in their livers than did livers of lean rats. cAMP levels were decreased in obese rats, which might explain the elevations of the other factors involved in carbohydrate metabolism.

Other tissues in which fructose-2,6-biphosphate has been studied include intestine,[91] muscle,[91] kidney,[92] and pancreatic islets.[93,94] Mice,[91] swine,[92] hamsters,[95] and yeasts,[96,97] as well as rats, have been used to study the metabolism of fructose-2,6-biphosphate.

Mizunuma and Tashima[91] compared the effects of Mn^{++} and Mg^{++} on the inhibition of fructose-1,6-diphosphatase from the liver, intestine, and muscle of mice by fructose-2,6-biphosphate. The enzymes from all three tissues were inhibited by fructose-2,6-biphosphate. The inhibition of the liver enzyme by fructose-2,6-biphosphate was much greater in the presence of Mg^{++} than of Mn^{++}. The effect of Mn^{++} may be due to a desensitization of AMP inhibition of the enzyme. The extent of the release of inhibition of fructose-1-phosphatase by fructose-2,6-biphosphate and AMP due to Mn^{++} was decreased in the order of the liver, muscle, and intestine isoenzymes. It was concluded that fructose-2,6-biphosphate apparently was not the major regulator of fructose-1,6-diphosphatase activity in skeletal muscle.

Fructose-2,6-biphosphate has also been reported to regulate carbohydrate metabolism in swine kidney.[92] Fructose-2,6-biphosphate decreased significantly during gluconeogenesis, as did fructose-1,6-diphosphate. Fructose-2,6-biphosphate in swine kidney also inhibited the activity of fructose-1,6-diphosphatase and stimulated 6-phosphofructo-1-kinase activity.

In pancreatic islets, fructose-2,6-biphosphate also acts as an activator of 6-phosphofructo-1-kinase.[93] Pancreatic islets were isolated from fed rats and then incubated with or without glucose. In the islets incubated with glucose, fructose-2,6-biphosphate was higher than those incubated without glucose, but glucose did not affect the activity of 6-phosphofructo-2-kinase. It was concluded that fructose-2,6-biphosphate increased as a result of the increase in fructose-6-phosphate of glucose-incubated islets.

Another study of fructose-2,6-biphosphate in pancreatic islets used cells from rats perfused with physiologic concentrations of glucose.[94] Modest increases of fructose-2,6-biphosphate occurred as the glucose concentration increased from 5.5 to 10 to 16.7 mM, but did not increase further at higher glucose concentrations. Levels of fructose-2,6-biphosphate were only about 10% of levels measured in liver, in contrast to reports by Malaisse et al.,[93] who found similar levels in liver and islets. In islets glucagon does not affect fructose-2,6-biphosphate as it does in the liver.[94]

Fructose-2,6-biphosphate has also been studied in yeast.[96,97] Fructose-1,6-diphosphatase isolated from various strains of Saccharomyces cerevisiae was incubated with cAMP-dependent protein kinase and fructose-2,6-biphosphate. The inhibitory effect of fructose-2,6-biphosphate on fructose-1,6-diphosphatase was similar to effects found in liver and other tissues, demonstrating that this fructose metabolite is an important factor in the regulation of carbohydrate metabolism in single-celled organisms as well as mammalian tissues. Research of the effects of this compound on carbohydrate metabolism is very recent, and many details concerning its role need to be investigated further.

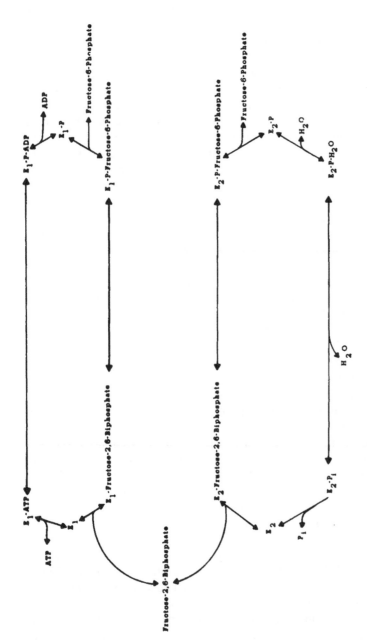

FIGURE 8.. Proposed two-site model of reaction mechanisms of 6-phosphofructo-2-kinase/fructose-2,6-biphosphatase. (Adapted from Pilkis, S. J., Regen, D. M., Stewart, H. B., Pilkis, J., Pate, T. M., and El-Maghrabi, M. R., *J. Biol. Chem.*, 259, 949, 1984.)

REFERENCES

1. **Mendeloff, A. I. and Weichselbaum, T. E.,** Role of the human liver in the assimilation of intravenously administered fructose, *Metab. Clin. Exp.,* 2, 450, 1953.
2. **Levine, R. and Huddleston, B.,** The comparative action of insulin on the disposal of intravenous fructose and glucose, *Fed. Proc., Fed. Am. Soc. Exp. Biol.,* 6, 151, 1947.
3. **Bollman, J. L. and Mann, F. C.,** The physiology of the liver; the utilization of fructose following complete removal of liver, *Am. J. Physiol.,* 96, 683, 1931.
4. **Reinecke, R. M.,** The kidney as a locus of fructose metabolism, *Am. J. Physiol.,* 141, 669, 1944.
5. **Wick, A. N., Sherill, J. W., and Drury, D. R.,** The metabolism of fructose by the extrahepatic tissues, *Diabetes,* 2, 465, 1953.
6. **Weichselbaum, T. E., Margraf, H. W., and Elman, R.,** Metabolism of intravenously infused fructose in man, *Metab. Clin. Exp.,* 2, 434, 1953.
7. **Mackler, B. and Guest, G. M.,** Effects of insulin and glucose on utilization of fructose by isolated rat diaphragm, *Proc. Soc. Exp. Biol. Med.,* 83, 327, 1953.
8. **Nakada, H. I.,** The metabolism of fructose by isolated rat diaphragms, *J. Biol. Chem.,* 219, 319, 1956.
9. **Sols, A. and Crane, R. K.,** Substrate specificity of brain hexokinase, *J. Biol. Chem.,* 210, 581, 1954.
10. **Vinuela, E., Salas, M., and Sols, A.,** Glucokinase and hexokinase in liver in relation to glycogen synthesis, *J. Biol. Chem.,* 238, PC1175, 1963.
11. **Wang, Y.-M. and van Eys, J.,** Nutritional significance of fructose and sugar alcohols, *Annu. Rev. Nutr.,* 1, 437, 1981.
12. **Sestoft, L. and Fleron, P.,** Determination of the kinetic constants of the fructose transport and phosphorylation in the perfused rat liver, *Biochim. Biophys. Acta,* 345, 27, 1974.
13. **Herman, R. H.,** Hydrolysis and absorption of carbohydrates and adaptive response of the jejunum, in *Sugars in Nutrition,* Sipple, H. L. and McNutt, K. W., Eds., Academic Press, Orlando, Fla., 1974, 145.
14. **Bode, C., Eisenhardt, J. M., Haberich, F. J., and Bode, J. C.,** Influence of feeding fructose on fructose and glucose absorption in rat jejunum and ileum, *Res. Exp. Med.,* 179, 163, 1981.
15. **Leuthardt, F. and Testa, E.,** Die Phosphorylierung der Fructose in der Leber. II. Mitteilung, *Helv. Chim. Acta,* 34, 931, 1951.
16. **Hers, H.-G.,** La fructokinase du foie, *Biochim. Biophys. Acta,* 8, 416, 1952.
17. **Cori, G. T., Ochoa, D., Slein, M. W., and Cori, C. F.,** The metabolism of fructose in the liver. Isolation of fructose-1-phosphate and inorganic pyrophosphate, *Biochim. Biophys. Acta,* 7, 304, 1951.
18. **Sanchez, J. J., Gonzalez, N. S., and Pontis, H. G.,** Fructokinase from rat liver. I. Purification and properties, *Biochim. Biophys. Acta,* 227, 67, 1971.
19. **Adelman, R. C., Ballard, F. J., and Weinhouse, S.,** Purification and properties of rat liver fructokinase, *J. Biol. Chem.,* 242, 3360, 1967.
20. **Sanchez, J. J., Gonzalez, N. S., and Pontis, H. G.,** Fructokinase from rat liver. II. The role of K^+ on the enzyme activity, *Biochim. Biophys. Acta,* 227, 79, 1971.
21. **Parks, R. E., Ben-Gershom, E., and Lardy, H. A.,** Liver fructokinase, *J. Biol. Chem.,* 227, 231, 1957.
22. **Leuthardt, F., Testa, E., and Wolf, H. P.,** Der enzymatische Abbau der Fructose 1-phosphate in der Leber. III. Mitteilung über den Stoffwechsel der Fructose in der Leber, *Helv. Chim. Acta,* 36, 227, 1953.
23. **Hers, H.-G. and Kusaka, T.,** Le métabolisme du fructose-1-phosphate dans le foie, *Biochim. Biophys. Acta,* 11, 427, 1953.
24. **Hers, H.-G. and Jacques, P. J.,** Hétérogénéité des aldolases, *Arch. Int. Physiol.,* 61, 260, 1953.
25. **Penhoet, E. E., Kochman, M., and Rutter, W. J.,** Isolation of fructose diphosphate aldolases A, B, and C, *Biochemistry,* 8, 4391, 1969.
26. **Penhoet, E. E., Kochman, M., and Rutter, W. J.,** Molecular and catalytic properties of aldolase C, *Biochemistry,* 8, 4396, 1969.
27. **Hers, H.-G.,** The conversion of fructose 1-C^{14} and sorbitol 1-C^{14} to liver and muscle glycogen in the rat, *J. Biol. Chem.,* 214, 373, 1955.
28. **Rauschenbach, P. and Lamprecht, W.,** Einbau von ^{14}C-markierter Glucose and Fructose in Leberglykogen. Zum Fructose-Stoffwechsel in der Leber, *Hoppe Seyler's Z. Physiol. Chem.,* 339, 277, 1964.
29. **Landau, B. R. and Merlevede, W.,** Initial reactions in the metabolism of D- and L-glyceraldehyde by rat liver, *J. Biol. Chem.,* 238, 861, 1963.
30. **Förster, H.,** Comparative metabolism of xylitol, sorbitol, and fructose, in *Sugars in Nutrition,* Sipple, H. L. and McNutt, K. W., Eds., Academic Press, Orlando, Fla., 1974, 259.
31. **Frandsen, E. K. and Grunnet, N.,** Kinetic properties of triokinase from rat liver, *Eur. J. Biochem.,* 23, 588, 1971.
32. **Heinz, F. and Lemprecht, W.,** Anreicherung und Charakterisierung einer Triosekinase aus Leber. Zur Biochemie des Fructosestoffwechsels. III, *Hoppe Seyler's Z. Physiol. Chem.,* 324, 88, 1961.
33. **Hers, H.-G.,** Triokinase, in *Methods in Enzymology,* Vol. 5, Colowick, S. P. and Kaplan, N. O., Eds., Academic Press, Orlando, Fla., 1962, 362.

34. **Heinz, F.**, Enzyme des Fructosestoffwechsels: Änderungen von Enzymaktivitaten in Leber und Niere der Ratte bei Fructose und Glucose reicher Ernahrung, *Hoppe Seyler's Z. Physiol. Chem.*, 349, 399, 1968.
35. **Veneziale, C. M.**, Regulation of D-triokinase and NAD-linked glycerol-dehydrogenase activities in rat liver, *Eur. J. Biochem.*, 31, 59, 1972.
36. **Lamprecht, W. and Heinz, F.**, Isolierung von Glycerinaldehyd-Dehydrogenase aus Rattenleber. Zur Biochemie des Fructosestoffwechsels, *Z. Naturforsch Teil B*, 13, 464, 1958.
37. **Holzer, H. and Holldorf, A.**, Anreicherung, Charakterisierung und biologische Bedeutung einer D-Glycerat-kinase aus Rattenleber, *Biochem. Z.*, 329, 283, 1957.
38. **Ichihara, A. and Greenberg, D. M.**, Studies on the purification and properties of D-glyceric acid kinase of liver, *J. Biol. Chem.*, 225, 949, 1957.
39. **Lamprecht, W., Diamantstein, T., Heinz, F., and Balde, P.**, Phosphoryleirung von D-Glycerinsaure zu 2-Phospho-D-Glycerinsaure mit Glyceratkinase in Leber. I. Zur Biochemie des Fructosestoffwechsels. II, *Hoppe Seyler's Z. Physiol. Chem.*, 316, 97, 1959.
40. **Kattermann, R. and Holzer, H.**, D-Glycerate beim Fructoseabbau in der Leber, *Biochem. Z.*, 334, 218, 1961.
41. **Sillero, M. A. G., Sillero, A., and Sols, A.**, Enzymes involved in fructose metabolism in liver and the glyceraldehyde metabolic crossroads, *Eur. J. Biochem.*, 10, 345, 1969.
42. **Heinz, F.**, The enzymes of carbohydrate degradation, *Prog. Biochem. Pharmacol.*, 8, 1, 1973.
43. **Wolf, H. P. and Leuthardt, F.**, Über die Glycerindehydrase der Leber, *Helv. Chim. Acta*, 36, 1463, 1953.
44. **Hers, H.-G.**, L'aldose réductase, *Biochim. Biophys. Acta*, 37, 120, 1960.
45. **Bublitz, C. and Kennedy, E. P.**, A note on the asymmetrical metabolism of glycerol, *J. Biol. Chem.*, 211, 963, 1954.
46. **Burch, H. B., Max, P., Jr., Chyu, K., and Lowry, O. H.**, Metabolic intermediates in liver of rats given large amounts of fructose or dihydroxyacetone, *Biochem. Biophys. Res. Commun.*, 34, 619, 1969.
47. **Burch, H. B., Lowry, O. H., Meinhardt, L., Max, P., Jr., and Chyu, K.**, Effects of fructose, dihydroxyacetone, glycerol and glucose on metabolites and related compounds in liver and kidney, *J. Biol. Chem.*, 245, 2092, 1970.
48. **Günther, M. A., Sillero, A., and Sols, A.**, Fructokinase assay with a specific spectrophotometric method using 1-phosphofructokinase, *Enzymol. Biol. Clin.*, 8, 341, 1967.
49. **Heinz, F. and Junghänel, J.**, Metabolitmuster in Rattenleber nach Fructoseapplikation, *Hoppe Seyler's Z. Physiol. Chem.*, 350, 859, 1969.
50. **Mäenpää, P. H., Raivio, K. O., and Kekomaki, M. P.**, Liver adenine nucleotides: fructose induced depletion and its effect on protein synthesis, *Science*, 161, 1253, 1968.
51. **Raivio, K. O., Kekomaki, M. P., and Mäenpää, P. H.**, Depletion of liver adenine nucleotides induced by D-fructose, dose-dependence and specificity of the fructose effect, *Biochem. Pharmacol.*, 18, 2615, 1969.
52. **Bode, C., Schumacher, H., Goebell, H., Zelder, O., and Pelzel, H.**, Fructose induced depletion of liver adenine nucleotides in man, *Horm. Metab. Res.*, 3, 289, 1971.
53. **Sestoft, L.**, Effects of rapidly phosphorylated substrates fructose, sorbitol, xylitol (glycerol): toxic or metabolic toxicological aspect of food safety, *Arch. Toxicol.*, Suppl. 1, 151, 1978.
54. **Tuovinen, C. G. R. and Bender, A. E.**, Some metabolic effects of prolonged feeding of starch, sucrose, fructose and carbohydrate-free diet in rats, *Nutr. Metab.*, 19, 161, 1975.
55. **Morris, R. C., Jr., Nigon, K., and Reed, E. B.**, Evidence that the severity of depletion of inorganic phosphate determines the severity of the disturbance of adenine nucleotide metabolism in the liver and renal cortex of the fructose-loaded rat, *J. Clin. Invest.*, 61, 209, 1978.
56. **Kreusser, W., Ritz, E., and Boland, R.**, Phosphate-depletion, *Klin. Wochenschr.*, 58, 1, 1980.
57. **Goldblatt, P. J., Witschi, H., Freedman, M. A., Sullivan, R. J., and Shull, K. H.**, Some structural and functional consequences of hepatic adenosine triphosphate deficiency induced by intraperitoneal fructose administration, *Lab. Invest.*, 23, 378, 1970.
58. **Phillips, M. J., Heteny, G., Jr., and Adachi, F.**, Ultrastructural hepatocellular alterations induced by in vivo fructose infusion, *Lab. Invest.*, 22, 370, 1970.
59. **Shinozura, H., Reid, I. M., Shull, K. H., Liang, H., and Farber, E.**, Dynamics of liver cell injury and repair, *Lab. Invest.*, 23, 253, 1970.
60. **Förster, H., Meyer, E., Ziege, M.**, Erhöhung von Serumharnsäure und Serumbilirubin nach hochdosierten Infusionen von Sorbit, Xylit und Fructose, *Klin. Wochenschr.*, 48, 878, 1970.
61. **Hallfrisch, J., Ellwood, K., Michaelis, O. E., IV, Reiser, S., and Prather, E. S.**, Plasma fructose, uric acid, and inorganic phosphorus responses of hyperinsulinemic men fed fructose, *J. Am. Coll. Nutr.*, 5, 61, 1986.
62. **Israel, K. D., Michaelis, O. E., IV, Reiser, S., and Keeney, M.**, Serum uric acid, inorganic phosphorus, and glutamic-oxalacetic transaminase and blood pressure in carbohydrate-sensitive adults consuming three different levels of sucrose, *Ann. Nutr. Metab.*, 27, 425, 1983.

63. **Solyst, J. T., Michaelis, O. E., IV, Reiser, S., Ellwood, K. C., and Prather, E. S.,** Effect of dietary sucrose in humans on blood uric acid, phosphorus, fructose, and lactic acid responses to a sucrose load, *Nutr. Metab.*, 24, 182, 1980.
64. **Fox, I. H. and Kelley, W. N.,** Studies on the mechanism of fructose induced hyperuricemia in men, *Metab. Clin. Exp.*, 23, 713, 1974.
65. **Jacobs, D.,** Hyperuricemia as a risk factor in coronary heart disease, *Adv. Exp. Med. Biol.*, 76B, 231, 1977.
66. **Herman, J. B., Medalie, J. H., and Goldbourt, N.,** Diabetes, prediabetes and uricaemia, *Diabetologia*, 21, 47, 1976.
67. **Macdonald, I., Keyser, A., and Pacy, D.,** Some effects, in man, of varying the load of glucose, sucrose, fructose, or sorbitol on various metabolites in blood, *Am. J. Clin. Nutr.*, 31, 1305, 1978.
68. **Emmerson, B. T.,** Effect of oral fructose on urate production, *Ann. Rheum. Dis.*, 33, 276, 1974.
69. **Fox, I. H. and Kelley, W. N.,** Studies on the mechanism of fructose-induced hyperuricemia in man, *Metabolism*, 21, 713, 1972.
70. **Sahebjami, H. and Scalettar, R.,** Effects of fructose infusion on lactate and uric acid metabolism, *Lancet*, 1, 366, 1971.
71. **El-Ebrashy, N., Shaheen, M. H., Wasfi, A. A., and El-Danasoury, M.,** Side effects of IV fructose load in diabetics, *J. Egypt. Med. Assoc.*, 57, 406, 1974.
72. **Fiaschi, E., Baggio, B., Favaro, S., Antonello, A., Camerin, E., Todesco, S., and Borsatti, A.,** Fructose-induced hyperuricemia in essential hypertension, *Metabolism*, 26, 1219, 1977.
73. **Perheentupa, J., Raivio, K. O., and Nikkilä, E. A.,** Hereditary fructose intolerance, *Acta Med. Scand.*, Suppl. 542, 65, 1972.
74. **Huttunen, J. K., Makinen, K. K., and Scheinin, A.,** Turku sugar studies. XI. Effects of sucrose, fructose and xylitol diets on glucose, lipid, and urate metabolism, *Acta Odontol. Scand.*, 33, Suppl. 70, 239, 1975.
75. **Turner, J. L., Bierman, E. L., Brunzell, J. D., and Chait, A.,** Effect of dietary fructose on triglyceride transport and glucoregulatory hormones in hypertriglyceridemic men, *Am. J. Clin. Nutr.*, 32, 1043, 1979.
76. **Fox, I. H., John, D., DeBruyne, S., Dwosh, I., and Marliss, E. B.,** Hyperuricemia and hypertriglyceridemia: metabolic basis for the association, *Metabolism*, 34, 741, 1985.
77. **Claus, T. H., El-Maghrabi, M. R., Regen, D. M., Stewart, H. B., McGrane, M., Kountz, P. D., Nyfeler, F., Pilkis, J., and Pilkis, S. J.,** The role of fructose 2,6-biphosphate in the regulation of carbohydrate metabolism, *Curr. Top. Cell. Regul.*, 23, 57, 1984.
78. **Van Schaftingen, E. and Hers, H.-G.,** Synthesis of a stimulator of phosphofructokinase, most likely fructose 2,6-biphosphate, from phosphoric acid and fructose 6-phosphoric acid, *Biochem. Biophys. Res. Commun.*, 96, 1524, 1980.
79. **Pilkis, S. J., Walderhaug, M., Murray, K., Beth, A., Venkataramu, S. D., Pilkis, J., and El-Maghrabi, M. R.,** 6-Phosphofructo-2-kinase/fructose 2,6-diphosphatase from rat liver. Isolation and identification of a phosphorylated intermediate, *J. Biol. Chem.*, 258, 6135, 1983.
80. **Pilkis, S. J., Chrisman, T. D., El-Maghrabi, M. R., Colosia, A., Fox, E., Pilkis, J., and Claus, T. H.,** The action of insulin on hepatic fructose 2,6-biphosphate metabolism, *J. Biol. Chem.*, 258, 1495, 1983.
81. **El-Maghrabi, M. R., Claus, T. H., Pilkis, J., Fox, E., and Pilkis, S. J.,** Regulation of rat liver fructose 2,6-biphosphatase, *J. Biol. Chem.*, 257, 7603, 1982.
82. **Miernyk, J. A. and Dennis, D. T.,** Activation of the plastid isozyme of phosphofructokinase from developing endosperm of *Ricinus communis* by fructose 2,6-biphosphate, *Biochem. Biophys. Res. Commun.*, 105, 793, 1982.
83. **Richards, C. S., Yokoyama, M., Furuya, E., and Uyeda, K.,** Reciprocal changes in fructose-6-phosphate,2-kinase and fructose-2,6-biphosphate activity in response to glucagon and epinephrine, *Biochem. Biophys. Res. Commun.*, 104, 1073, 1982.
84. **Hue, L.,** Role of fructose-2,6-biphosphate in the stimulation of glycolysis by anoxia in isolated hepatocytes, *Biochem. J.*, 206, 359, 1982.
85. **Neely, P., El-Maghrabi, M. R., Pilkis, S. J., and Claus, T. H.,** Effect of diabetes, insulin, starvation, and refeeding on the level of rat hepatic fructose 2,6-biphosphate, *Diabetes*, 30, 1062, 1981.
86. **Bartrons, R., Hue, L., Van Schaftingen, E., and Hers, H.-G.,** Hormonal control of fructose 2,6-biphosphate concentration in isolated rat hepatocytes, *Biochem. J.*, 214, 829, 1983.
87. **El-Maghrabi, M. R., Claus, T. H., Pilkis, J., and Pilkis, S. J.,** Regulation of 6-phosphofructo-2-kinase activity by cyclic AMP-dependent phosphorylation, *Proc. Natl. Acad. Sci. U.S.A.*, 79, 315, 1982.
88. **Reinhart, G. D.,** Influence of fructose 2,6-bisphosphate on the aggregation properties of rat liver phosphofructokinase, *J. Biol. Chem.*, 258, 10827, 1983.
89. **Pilkis, S. J., Regen, D. M., Stewart, H. B., Pilkis, J., Pate, T. M., and El-Maghrabi, M. R.,** Evidence for two catalytic sites on 6-phosphofructo-2-kinase/fructose 2,6-biphosphatase. Dynamics of substrate exchange and phosphoryl enzyme formation, *J. Biol. Chem.*, 259, 949, 1984.

90. **Hue, L., Van de Werve, G., and Jeanrenaud, B.,** Fructose-2,6-biphosphate in livers of genetically obese rats, *Biochem. J.,* 214, 1019, 1983.
91. **Mizunuma, H. and Tashima, Y.,** Effect of Mn^{2+} on fructose 2,6-biphosphate inhibition of mouse liver, intestinal, and muscle fructose-1,6-biphosphatases, *Arch. Biochem. Biophys.,* 226, 257, 1983.
92. **Muniyappa, K., Leibach, F. H., and Mendicino, J.,** Reciprocal regulation of 2,6-biphosphate in swine kidney, *Life Sci.,* 32, 271, 1983.
93. **Malaisse, W. J., Malaisse-Lagae, F., and Sener, A.,** Glucose-induced accumulation of fructose-2,6-biphosphate in pancreatic islets, *Diabetes,* 31, 90, 1982.
94. **Burch, P. T., Berner, D. K., Najafi, H., Meglasson, M. D., and Matschinsky, F. M.,** Regulatory role of fructose-2,6-biphosphate in pancreatic islet glucose metabolism remains unsettled, *Diabetes,* 34, 1014, 1985.
95. **Mizunuma, H. and Tashima, Y.,** Evidence for the intestinal type of fructose 1,6-diphosphatase in mouse, rat, and golden hamster, *Arch. Biochem. Biophys.,* 217, 512, 1982.
96. **Gancedo, J. M., Mazon, M. J., and Gancedo, C.,** Fructose 2,6-biphosphate activates the cAMP-dependent phosphorylation of yeast fructose-1,6-diphosphatase in vitro, *J. Biol. Chem.,* 258, 5998, 1983.
97. **Clifton, D. and Fraenkel, D. G.,** Fructose 2,6-biphosphate and fructose-6-P 2-kinase in *Saccharomyces cerevisiae* in relation to metabolic state in wild type and fructose-6-P 1-kinase mutant strains, *J. Biol. Chem.,* 258, 9245, 1983.

Chapter 4

INBORN ERRORS OF METABOLISM

I. INTRODUCTION

There are a few documented genetic errors of metabolism involving fructose. These errors involve either an absence or an abnormal production of an enzyme necessary for the utilization of fructose. Four enzyme defects of fructose metabolism which have been reported involve fructose-1-phosphate aldolase (E.C.4.1.2.7), fructokinase (E.C.2.7.1.4), fructose-1,6-diphosphatase (E.C.3.1.3.11), and phosphofructokinase (E.C.2.7.1.11). The absence or deficiency of these enzymes results in the buildup of metabolites which may have very damaging effects on the surrounding tissues.

II. HEREDITARY FRUCTOSE INTOLERANCE

Fructose-1-phosphate aldolase deficiency (hereditary fructose intolerance) has been described by a number of investigators.[1-26] This defect is probably the most serious of the genetic errors of fructose metabolism that will be described. Fructose-1-phosphate aldolase catalyzes the conversion of fructose-1-phosphate to D-glyceraldehyde and dihydroxyacetone phosphate (Figure 1). Its deficiency results in the buildup of fructose-1-phosphate. Symptoms of this disease include vomiting after fructose ingestion, hypoglycemia, and elevated fructose in blood and urine. In children there are growth retardation, jaundice, elevated bilirubin, and urinary excretion of albumin and amino acids. Death has been reported after fructose infusions. Summaries of acute and chronic symptoms appear in Tables 1 and 2. Acute symptoms are those which appear 20 min after oral consumption of fructose. Chronic symptoms usually appear after weaning when foods containing fructose or sucrose are added to the diet.

In 1956 the first case of hereditary fructose intolerance was reported in a 24-year-old woman. The young woman's only symptoms were "feelings of anxiety" and nausea and vomiting after eating fruit or food containing fructose. Chambers and Pratt[1] called this syndrome "idiosyncracy to fructose". The next year Froesch et at.[2] reported similar symptoms in two siblings and two other relatives. In these four patients blood fructose levels ranged from 145 to 160 mg/dℓ. Normal peak levels after fructose ingestion are less than 50 mg/dℓ. Urinary protein and amino acid levels were increased after fructose ingestion. Froesch et al. were the first to recognize the genetic heritability of the disease. Since the discovery of this defect, over 100 cases in 7 European countries and the U.S. have been studied.[1-26]

Hers and Joassin[3] reported that the activity of fructose-1,6-diphosphate aldolase in the liver of patients with fructose intolerance was only 25% of normal. They postulated that the reduction of this enzyme could be due to the lack of fructose-1-phosphate aldolase, which would result in an accumulation of fructose-1-phosphate and thereby competitively inhibit the binding of fructose-1,6-diphosphate to the fructose-1,6-diphosphate aldolase. In normal subjects the ratio of fructose-1,6-diphosphate aldolase to fructose-1-phosphate aldolase is 1 to 1.5,[5] but in patients with hereditary fructose intolerance this ratio is much higher, ranging from 3 to 11. Morris et al.[6] found not only liver, but also renal fructose-1-phosphate aldolase activity reduced. The ratio of the two aldolases was 5 compared to 1 to 1.5 in kidneys of normal humans. Normal fructose-1,6-diphosphate aldolase activity with deficient fructose-1-phosphate aldolase activity was found in the jejunal mucosa of one patient.[7] Kranhold et al.[8] studied livers and various organs and reported finding fructose-1-phosphate aldolase in liver, kidney, and small intestine. These organs would then be subject to accumulation of

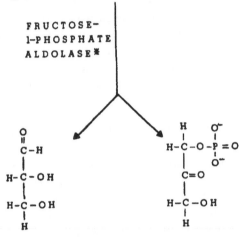

α-D-FRUCTOSE-1-PHOSPHATE

FRUCTOSE-
1-PHOSPHATE
ALDOLASE*

D-GLYCERALDEHYDE   DIHYDROXYACETONE
                          PHOSPHATE

FIGURE 1.   Metabolic defect resulting in hereditary fructose intolerance.
(*) Deficiency or absence of this enzyme results in accumulation of fruc-
tose-1-phosphate.

Table 1
ACUTE SYMPTOMS OF HEREDITARY
FRUCTOSE INTOLERANCE

Aminoaciduria	Nausea
Dizziness	Elevated hepatic enzyme levels
Fructosemia (moderate)	Sweating
Fructosuria	Trembling
Hyperbilirubinemia	Vomiting
Hypoglycemia	Loss of consciousness
Hypophosphatemia	Coma

Table 2
CHRONIC SYMPTOMS OF
HEREDITARY FRUCTOSE
INTOLERANCE

Failure to thrive	Aversion to sweets
Vomiting	Lack of dental caries
Dehydration	Fructosemia
Hepatomegaly	Fructosuria
Jaundice	Hyperbilirubinemia
Edema	Hypophosphatemia
Seizures	Elevated hepatic enzyme levels
Ascites	Liver cirrhosis
Hyperuricosuria	Hyperuricemia

fructose-1-phosphate in hereditary fructose intolerance and result in dysfunction and degeneration of these tissues.

Early diagnosis is important in order to prevent organ damage and even death. Firstborn children with the disease suffer more serious damages and are more likely to die than subsequent siblings.[5] Older children and adults often exhibit no symptoms because they have developed a natural aversion to any sweet foods. Infants have much less control over intake and may just stop eating to avoid the unpleasant effects. The ensuing cachexia may result in death.[5] Elevations in liver glycogen stores have resulted in some patients being misdiagnosed as having glycogen storage disease.[9] Hereditary fructose intolerance may appear at weaning when fructose-containing foods are introduced. There is a failure to thrive, in addition to vomiting and hypoglycemia; unconsciousness may occasionally occur, as well as jaundice, enlarged spleen and liver, and excessive urinary excretion of albumin and amino acids. Black and Simpson[10] reported a 6-month-old infant to weigh only about 1 lb above its birth weight of about 6 lb when hereditary fructose intolerance was diagnosed. The norm for this child would have been 13 to 14 lb. After diagnosis and introduction of a fructose-free diet, the child rapidly caught up, and by 9 months of age was normal weight. Levin et al.[11] reported edema and ascites in an infant $1\frac{1}{2}$ months old at diagnosis of hereditary fructose intolerance. After only 2 weeks of a fructose-free diet, both of these symptoms disappeared.

Liver damage is reflected in increased levels of serum glutamic oxaloacetic transaminase and glutamic pyruvic transaminase activities.[11] Elevations of both of these enzymes are symptoms of hereditary fructose intolerance (Table 1). Lesions similar to early-stage cirrhosis were found in two children with hereditary fructose intolerance. Some of the more serious signs of liver damage which may occur include gross hepatomegaly, abdominal distention with ascites, and hypoalbuminemia. Deficiency of hepatic coagulation factors is common and may result in GI or cerebral bleeding.[12] Hypochromic anemia is a common symptom.[21]

Hypoglycemia, which is responsible for the unconsciousness or even coma seen in some patients with hereditary fructose intolerance, is unresponsive to glucagon[11] and results from inhibition of glycogen breakdown by accumulated fructose-1-phosphate.[3,6,7] Hypoglycemia occurs whether fructose is eaten or infused and does not respond to elevated blood glucagon levels.[14] The hypoglycemia is not the result of hyperinsulinemia,[5] but appears to be the result of inhibited release of glucose from the liver. When fructose was given, insulin was either unchanged or less than fasting during fructose-induced hypoglycemia.[4,5] This may result from depletion of inorganic phosphorus and subsequent reduction of ATP in the liver, which may lead to inactivation of glycogen phosphorylase and therefore prevent conversion of glycogen to glucose. Decreased availability of phosphorus would result in higher degradation of nucleotides to uric acid, and in hyperuricemia and hyperuricosuria.[15-20]

Fructose-1-phosphate and fructose-1,6-diphosphate aldolase activities were measured in both normal and hereditary fructose-intolerant patients.[4,5,10,22-26] A summary of these results is listed in Table 3. Schmidt et al.[22] found enzyme levels of normal patients to be 229 ± 107 U/g for fructose-1,6-diphosphate aldolase and 164 ± 76 U/g for fructose-1-phosphate aldolase. The ratio of fructose-1,6-diphosphate aldolase to fructose-1-phosphate aldolase was 1.4. In patients with hereditary fructose intolerance, although fructose-1,6-diphosphate aldolase levels varied widely from 2.7 to the almost normal level of 200, fructose-1-phosphate aldolase was undetectable or substantially reduced. Therefore, the ratio of the two enzymes increased substantially, making this ratio a good marker for this disease.

Froesch et al.[4] measured fructose-1,6-diphosphate aldolase levels of two adults with hereditary fructose intolerance, and although their levels were 125 and 200 U/g, the ratio of activity of the diphosphate aldolase to that of the monophosphate aldolase was greater than 10 in both subjects. Most of the enzyme levels have been examined in infants. Both Froesch[5] and Nikkilä et al.[26] evaluated patients in whom they were unable to detect fructose-

Table 3
**FRUCTOSE-1-PHOSPHATE ALDOLASE (F1P) AND
FRUCTOSE-1,6-DIPHOSPHATE ALDOLASE (FDP) IN
THE LIVER OF NORMAL SUBJECTS AND PATIENTS
WITH HEREDITARY FRUCTOSE INTOLERANCE**

Subjects	FDP (U/g)	F1P (U/g)	Ratio (FDP/F1P)	Ref.
2 normal[a]	229 ± 107	164 ± 76	1.4	22
2 adults	125, 200	12, 19	10.4, 10.5	4
1 infant	50	ND[b]	—	5
1 infant	187	5	37.4	11
1 infant	137	22	6.2	23
4 infants	55, 63, 102, 55	5, 9, 16, 11	11.0, 6.7, 6.2, 5.0	24
2 infants	190, 83	28, 16	6.7, 5.2	25[c]
2 infants	6.5, 2.7	ND, 1.0	—, 2.7	26[d]

[a]   All other subjects were those with hereditary fructose intolerance.
[b]   ND = not detectable.
[c]   Studied two patients of Hers[24].
[d]   Enzyme activities expressed as μg triose ester per milligram liver per 15 min rather than Bucher units.

Table 4
**CHANGES IN BLOOD GLUCOSE, LACTATE, AND
URIC ACID CONCENTRATIONS AFTER A DOSE OF
0.4 g FRUCTOSE PER KILOGRAM BODY WEIGHT
FOR HEREDITARY FRUCTOSE-INTOLERANT (HF1)
CHILDREN AND 0.5 g FRUCTOSE PER KILOGRAM
BODY WEIGHT IN NORMAL CHILDREN**

	HF1 children ($n = 5$)		Normal children ($n = 5$)	
	Δ mg/dℓ	%	Δ mg/dℓ	%
Uric acid	+4.5	+115	+2.8	+92
Lactic acid	+21.5	+188	+22.6	+149
Glucose	−39	−55	+25	+39

Adapted from Perheentupa, J., Raivio, K. O., and Nikkilä, E. A., *Acta Med. Scand. Suppl.*, 542, 65, 1972.

1-phosphate aldolase. The ratios of the enzymes in the infants reported in these studies (Table 3) ranged from a low of 2.7 in one infant with levels of both enzymes at about 10% of normal[26] to a high of 37.4 in another infant with close to normal levels of fructose-1,6-diphosphate aldolase, but fructose-1-phosphate aldolase levels less than 5% of those in normal adults.[10] These studies confirm the primary enzyme defect to be deficiency or absence of fructose-1-phosphate aldolase in patients with hereditary fructose intolerance.

Perheentupa et al.[21] have reviewed the clinical symptoms and pathologic consequences of hereditary fructose intolerance. In Finland 18 patients from 14 different families had been diagnosed with the genetic defect. Changes in blood glucose, lactate, and uric acid concentration after a fructose dose of 0.4 or 0.5 g/kg body weight in five normal children and five children with hereditary fructose intolerance are summarized in Table 4. There was a greater increase in both uric and lactic acid in the children with hereditary fructose intolerance than in the normal controls, despite a lower dose (0.4 vs 0.5 g fructose per kilogram body weight). The patients also exhibited the typical decrease in blood glucose in response to fructose, while the normal children showed substantial increases in blood glucose.

Perheentupa and Raivio[27] noticed that patients with hereditary fructose intolerance had urate sediment in their urine. This observation led to the discovery that uric acid production was increased in these patients and also in normal patients after fructose ingestion. The increase in uric acid is a result of increased degradation of liver adenine nucleotides caused by the sequestering of inorganic phosphorus.[17,19,20,28] The accumulation of fructose-1-phosphate and concomitant decrease of inorganic phosphorus last longer and are more pronounced in fructose-intolerant patients than in normal subjects, and therefore hyperuricemia after fructose is also greater.[16,29] Neither phosphate nor glucose infusions prevent hyperuricemia in patients with hereditary fructose intolerance.[21]

Nausea and vomiting occur after eating fructose, but not necessarily after infusion. These symptoms indicate an enzyme defect in the intestine.[8] Glucose absorption may also be impaired. Accumulation of fructose-1-phosphate in the intestinal mucosa may be the cause of reduced glucose absorption.[30,31] Diarrhea, which has been reported in a few cases of hereditary fructose intolerance, may be a clinical manifestation of intestinal glucose malabsorption.[4,32,33]

In the kidneys of patients with hereditary fructose intolerance, symptoms similar to Fanconi's syndrome occur.[29,34,35] Some renal symptoms include cystinosis, renal calculi, and potassium depletion. Acute loads of fructose impair the ability of the proximal renal tubule to reabsorb amino acids, water, bicarbonate, and phosphate. The first sign of loss of renal function is the inability to acidify urine. This "renal tubular acidosis" disappears when fructose ingestion ends. Fructose ingestion rapidly causes aminoaciduria and proteinuria in patients with this defect. These pathologic effects on the kidney are a result of the accumulation of fructose-1-phosphate.

One unusual finding in patients with hereditary fructose intolerance is the lack of dental caries.[4,36,37] Levin et al.[36] examined the dental status of seven patients and Marthaler and Froesch[37] examined eight. Both groups of patients had a lower incidence of dental caries than would be expected. The aversion to sweets resulting from this disease provides evidence that a low-sugar diet does prevent tooth decay.

There seem to be no effects on intelligence in those older children and adults who survive with this disease.[2,4,11,36,38] Even though some patients have experienced frequent attacks of hypoglycemia, there have been no cases of serious brain damage reported in survivors. Growth retardation, however, can be severe.[15] A 5-year, 3-month-old boy was 100.1 cm tall (2nd percentile of the National Center for Health Statistics tables[39]); his weight was 15.2 kg (5th percentile). Although he had not been diagnosed to be fructose-intolerant, he had a history of aversion to sweets and fruits. His mother, noting that he "did not thrive on solid foods and infant formulas", had breastfed him until age 2 years, 4 months. Breast feeding alone for this length of time would have contributed to the growth retardation of this child. Diagnosis was made after an i.v. fructose tolerance test. Dietary records indicated that fructose intake prior to diagnosis averaged about 160 mg/kg body weight with a range of 110 to 450 mg/kg body weight. Stringent restriction to about 40 mg fructose per day resulted in an increase of growth velocity from the 25th to the 97th percentile. During the next 22 months, height of the patient increased from 2.71 S.D. below normal to 2.01 below.

A second growth-retarded child was diagnosed as fructose-intolerant at 6 months.[15] At 1 year of age, after a fructose-restricted diet his height had increased to 70 cm (1.3 percentile, 2.23 S.D. below normal). At 3 years he was 88.3 cm tall (3rd percentile, 1.88 S.D. below normal). After 3 years of age, growth velocity decreased. During this period dietary fructose intake was about 130 mg/kg body weight, and height for age decreased to 2.40 S.D. below normal or the 0.8 percentile. Fructose intake was restricted to 20 mg/kg body weight at age

4 years, 9 months, and for the next 9 months growth velocity doubled. In infants failure to thrive or retarded growth is one of the most common clinical symptoms reported for fructose intolerance. Although both boys[15] had severe growth retardation apparently as a result of chronic fructose intoxication, they did not experience the acute symptoms of nausea and vomiting after ingestion.

The genetic inheritability of this defect of fructose metabolism has been extensively studied by Froesch.[5] He has determined that this disorder results from an autosomal recessive defect, based on the findings that both sexes are affected and siblings are affected, while parents and children of patients are healthy. Fructose tolerance tests and fructose-1,6-diphosphate aldolase activities of five parents of hereditary fructose-intolerant children confirm this theory, showing parents to be normal heterogeneous carriers of a recessive trait.[40]

Early diagnosis of the disease is crucial in infants;[41] failure to diagnose can be fatal. Borrone et al.[42] examined eight infants with fructose intolerance. Two of these patients died, but the six survivors were reported to be "doing well on a restricted diet". Two young children with metabolic acidosis were treated with fructose infusion.[43] One patient, a 1-year-old boy, died after a 31-hr infusion of a solution containing 20% fructose. A second child, a 2-year-old girl, went into shock after a similar infusion. Luckily, the fructose infusion was stopped and the girl survived. Both of these patients had acidosis before the fructose infusions. Blood bicarbonate decreased during the fructose infusion, even though large doses of bicarbonate were concurrently administered. A previously undiagnosed $14^{1}/_{2}$-year-old boy also died after a 30-hr infusion of a total of 250 g fructose.[44] The boy had been admitted as a result of chronic abdominal pain, and an appendectomy was performed. Hereditary fructose intolerance was suspected at the end of the first day after the operation, and fructose infusion was stopped. The boy died 3 days later with acute kidney and liver failure. Diagnosis was confirmed by post-mortem examination of the liver. Wagner and Wolf[45] also reported a death involving fructose and sorbitol infusions.

These cases illustrate the importance of early diagnosis and the extreme danger of fructose infusions to fructose-intolerant individuals. Infants presenting any of the symptoms shown in Table 1, especially if just after weaning or when foods containing fructose have been introduced, should be placed on a fructose-free diet. Improvement or disappearance of symptoms would suggest this disorder. Diagnosis is usually confirmed by an i.v. fructose tolerance test. A 10% solution of 0.25 mg/kg body weight is given intravenously in 2 min. Blood inorganic phosphate and glucose are measured at 20, 40, 60, and 80 min. Both of these blood parameters decrease in hereditary fructose intolerance.[46] Steinmann and Gitzelmann[41] used both the fructose tolerance test and the aldolase analysis of liver biopsy to diagnose hereditary fructose intolerance. A standard fructose infusion of 200 mg/kg body weight was given to 11 children and 6 adults with hereditary fructose intolerance. Blood glucose, phosphorus, uric acid, magnesium, and fructose were measured at intervals for 2 hr. The fructose tolerance test was able to distinguish between the fructose-intolerant and 17 age-matched control children and 6 adult controls. Children with the defect had more severe and long-lasting hypoglycemia than adults. The fructose intolerance test can differentiate fructose intolerance from other acute liver diseases. Liver biopsies were taken from 35 fructose-intolerant children, 10 children suffering from other liver diseases, and 10 controls (9 organ donors). Although fructose aldolase was deficient in all fructose-intolerant children, it was also reduced in some children with other liver diseases. Other enzymes including fructose-1,6-diphosphatase and glucose-6-phosphatase needed to be examined to distinguish the fructose-intolerant children. Thus, the fructose tolerance test under controlled conditions in a hospital setting is the most useful diagnostic tool.

The only therapy for this disorder is a diet containing almost no fructose.[43-47] This diet would exclude all sucrose, many fruits, and vegetables. Mock et al.[15] used a diet containing 20 mg fructose per kilogram body weight per day to avoid growth retardation in children.

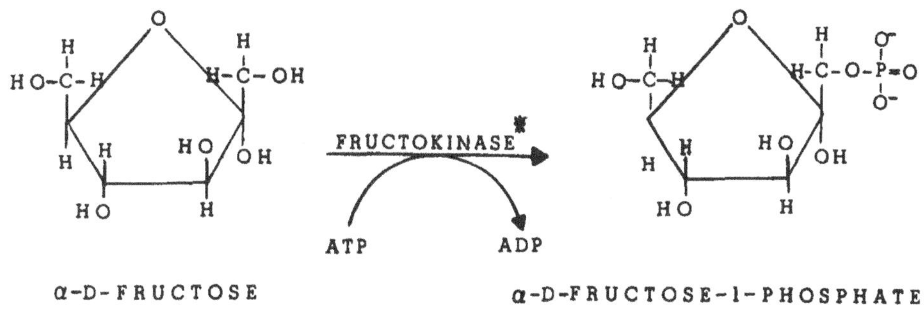

FIGURE 2. Metabolic defect resulting in essential fructosuria. (*) Deficiency or absence of this enzyme results in excretion of fructose in the urine after ingestion of fructose.

Table 5
DIAGNOSTIC
PARAMETERS OF
ESSENTIAL
FRUCTOSURIA AFTER
FRUCTOSE INGESTION

Elevated	No change
Blood fructose	Lactate
Urinary fructose	Phosphorus
	Glucose
	Pyruvate

This would amount to 1400 mg/day for a normal-sized adult man. One teaspoon of sugar contains approximately 2 g of fructose. Fructose has been used extensively in infusions and enteral and parenteral supplements. In view of the extreme danger of fructose to children with this defect, fructose infusions should be avoided if a patient is in shock, has fluid imbalances or metabolic acidosis, or has had surgery.

III. ESSENTIAL FRUCTOSURIA

Consequences of this defect are much less serious than hereditary fructose intolerance. In this case the enzyme defect is a lack of hepatic fructokinase.[25] Fructokinase is the enzyme responsible for phosphorylation of fructose in the liver to fructose-1-phosphate (Figure 2). Essential fructosuria is quite rare. Lasker[48] estimated incidence in the general population at 1 in 130,000. Approximately 60 cases were reviewed by Sachs et al;[49] about 25% were Jewish. At the Joslin Clinic only 4 of 29,000 diabetics were found to have the disease.[50] Since the disease is benign and asymptomatic, the incidence may be higher than reported. It usually is discovered during routine urinalysis for other conditions.[51]

Fructosuria occurs only after consumption of fructose-containing foods. During fructose tolerance tests, blood fructose rises to above 25 mg/dℓ and may reach 100 mg/dℓ. Blood fructose levels return to base line more slowly than in normal subjects.[51,52] These patients excrete 10 to 20% of fructose in the urine; in normal subjects only 1 to 2% is excreted.[5] There is no hypoglycemia or reduction in phosphorus, distinguishing essential fructosuria from hereditary fructose intolerance. Biochemical changes are summarized in Table 5. Patients with essential fructosuria respond to fructose ingestion with a smaller-than-normal increase in respiratory quotient,[53,54] indicating a reduced utilization of fructose. Lactate and pyruvate normally increase after fructose ingestion, but in these patients there is no change.[54,55]

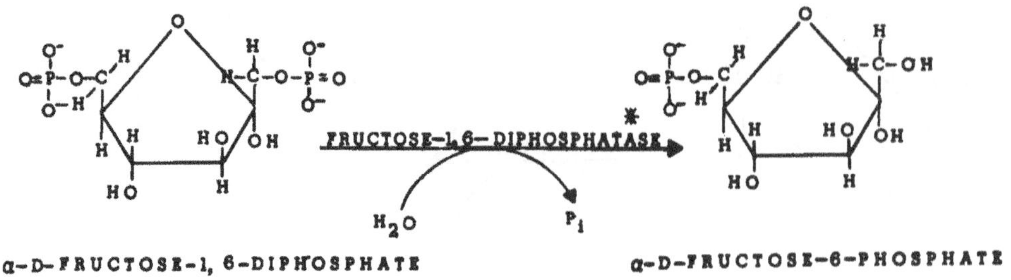

α-D-FRUCTOSE-1, 6-DIPHOSPHATE   α-D-FRUCTOSE-6-PHOSPHATE

FIGURE 3.   Metabolic defect resulting in hereditary fructose-1,6-diphosphatase deficiency. (*) Deficiency or absence of this enzyme results in accumulation of fructose-1,6-diphosphate.

Table 6
SYMPTOMS AND BIOCHEMICAL
CHANGES OCCURRING IN
HEREDITARY FRUCTOSE-1,6-
DIPHOSPHATASE DEFICIENCY[a]

Hepatomegaly	Poor muscle tone
Fasting hypoglycemia	Elevated blood lactate
Metabolic acidosis	

[a]   Growth and oral glucose and galactose tolerance are normal.

Lasker[48] has demonstrated that the defect is due to an inherited autosomal recessive trait. She studied 19 families in which more than 1 sibling had the defect. Of 15 families in which data on siblings were complete, 40 brothers and sisters of the original patients were tested; 7 of these patients had fructosuria. None of the parents or offspring of patients had the defect.

Since essential fructosuria is harmless and asymptomatic, there is no need for treatment.

IV. HEREDITARY FRUCTOSE-1,6-DIPHOSPHATASE DEFICIENCY

Very little information is available on this genetic defect of fructose metabolism. A deficiency of fructose-1,6-diphosphatase, which removes phosphate from the 1-position of fructose-1,6-diphosphate, is responsible for this hereditary defect (Figure 3). Fructose-1,6-diphosphatase deficiency has been reported by a few investigators.[56-58] Symptoms which are summarized in Table 6 include poor muscle tone, increased blood lactate levels, hepatomegaly, and fasting hypoglycemia which occurs spontaneously as a result of this disease or fructose-induced hypoglycemia. This disease is less severe than hereditary fructose intolerance, but can result in metabolic acidosis. Children do not vomit after fructose ingestion, and grow normally even when eating sucrose-containing foods. Baker and Winegrad[56] reported this disorder in a child of healthy parents whose first child had died. After meals there was a normal rise in glucose, but glucose dropped precipitously after fasting and did not respond to glucagon. It was concluded that glycogen storage was normal after a meal, but gluconeogenesis was deficient. In the liver the enzyme is deficient, but it appears to be normal in the muscle.[59] The fructose-1,6-diphosphatase in the muscle of the rabbit has been reported to be genetically distinct from the enzyme in the liver.[61] Fructose-1,6-disphophatase deficiency is also thought to be inherited via an autosomal recessive gene.

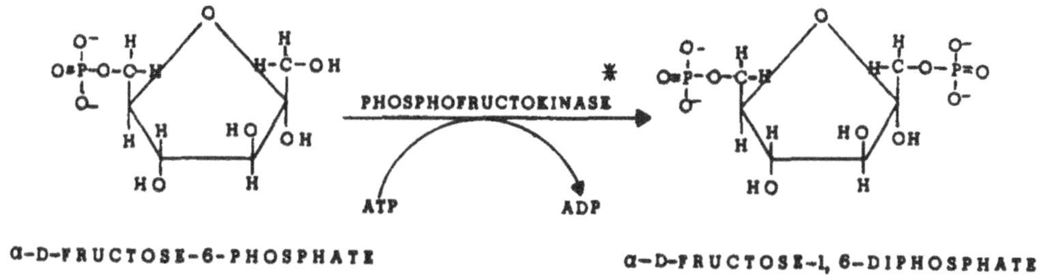

FIGURE 4.   Metabolic defect resulting in Type VII glycogenolysis. (*) Deficiency of this enzyme results in accumulation of fructose-6-phosphate in muscle.

Table 7
SYMPTOMS OF TYPE VII GLYCOGENOLYSIS

Fatigue   Muscle glycogen increase
Intolerance to exercise   Accumulation of hexose phosphate

V. TYPE VII GLYCOGENOLYSIS

Muscle phosphofructokinase deficiency has only been reported in a few patients.[61-64] This deficiency is more a defect of glucose metabolism than of fructose metabolism. There is a deficiency of phosphofructokinase which phosphorylates fructose-6-phosphate, a metabolite of glucose (Figure 4). This deficiency results in fatigue of subjects when they exercise, because hexose phosphate cannot be metabolized to produce energy. There is a reduction in the formation of ATP. This deficiency is also probably due to an autosomal recessive trait because it was reported in siblings of apparently unaffected parents.[62,63] Symptoms are listed in Table 7.

VI. SUMMARY

There are four known enzyme defects involving fructose metabolism. These are all considered to be the result of an inherited autosomal recessive gene resulting in the absence or deficiency of one of the following enzymes: fructose-1-phosphate aldolase, fructokinase, fructose-1,6-diphosphatase, or phosphofructokinase. The most serious and prevalent of these defects appears to be hereditary fructose intolerance, a deficiency of fructose-1-phosphate aldolase which, if undiagnosed, can and does result in death. Essential fructosuria, a relatively harmless and rare deficiency of fructokinase, results in increased excretion of fructose in the urine and is usually detected during routine analysis of urine. Fructose-1,6-diphosphatase deficiency results in hypoglycemia and hepatomegaly, but is much less severe than hereditary fructose intolerance, although it can result in acidosis. The final enzyme defect is a deficiency of phosphofructokinase in the muscle. This defect has only been reported in a few subjects. It results in fatigue after exercise and accumulation of hexose phosphates. It should probably be classified as a defect in glucose metabolism, since fructose-6-phosphate is for the most part a metabolite of glucose.

REFERENCES

1. **Chambers, R. A. and Pratt, R. T. C.**, Idiosyncracy to fructose, *Lancet*, 2, 340, 1956.
2. **Froesch, E. R., Prader, A., Labhart, A., Stuber, H. W., and Wolf, H. P.**, Die hereditare Fructosein-toleranz, eine bisher nicht bekannte kongenitale Stoffwechselstorung, *Schweiz. Med. Wochenschr.*, 87, 1168, 1957.
3. **Hers, H.-G. and Joassin, G.**, Anomalie de l'aldolase hépatique dans l'intolérance au fructose, *Enzymol. Biol. Clin.*, 1, 4, 1961.
4. **Froesch. E. R., Wolf, H. P., Baitsch, H., Prader, A., and Labhart, A.**, Hereditary fructose intolerance: an inborn defect of hepatic fructose-1-phosphate splitting aldolase, *Am. J. Med.*, 34, 151, 1963.
5. **Froesch, E. R.**, Essential fructosuria and hereditary fructose intolerance, in *The Metabolic Basis of Inherited Disease*, 3rd ed., Stanbury, J. B., Wyngaarden, J. B., and Fredrickson, D. S., Eds., McGraw-Hill, New York, 1972, 131.
6. **Morris, R. C., Jr., Ueki, I., Loh, D., Eanes, R. Z., and McLin, P.**, Absence of renal fructose-1-phosphate aldolase activity in hereditary fructose intolerance, *Nature*, 214, 920, 1967.
7. **Nisell, J. and Linden, L.**, Fructose-1-phosphate aldolase and fructose-1,6-diphosphate aldolase activity in the mucosa of the intestine in hereditary fructose intolerance, *Scand. J. Gastroenterol.*, 3, 80, 1968.
8. **Kranhold, J. F., Loh, D., and Morris, R. C., Jr.**, Renal fructose-metabolizing enzymes: significance in hereditary fructose intolerance, *Science*, 165, 402, 1969.
9. **Cain, A. R. R. and Ryman, B. E.**, High liver glycogen in hereditary fructose intolerance, *Gut*, 12, 929, 1971.
10. **Black, J. A. and Simpson, K.**, Fructose intolerance, *Br. Med. J.*, 4, 138, 1967.
11. **Levin, B., Oberholzer, V. G., Snodgrass, G. J. A. I., Stimmler, L., and Wilmers, M. J.**, Fructosemia an inborn error of fructose metabolism, *Arch. Dis. Child.*, 38, 220, 1963.
12. **Raju, L., Chessells, J. M., and Kemball, M.**, Manifestation of hereditary fructose intolerance, *Br. Med. J.*, 2, 446, 1971.
13. **Hers, H.-G.**, Inborn errors of carbohydrate metabolism, in *Sugars in Nutrition*, Sipple, H. L. and McNutt, K. W., Eds., Academic Press, Orlando, Fla., 1974, 337.
14. **Van den Berghe, G., Hue, L., and Hers, H.-G.**, Effect of the administration of fructose on the glyco-genolytic action of glucagon. An investigation of the pathogeny of hereditary fructose intolerance, *Biochem. J.*, 134, 637, 1973.
15. **Mock, D. M., Perman, J. A., Thaler, M. M., and Morris, R. C., Jr.**, Chronic fructose intoxication after infancy in children with hereditary fructose intolerance, *N. Engl. J. Med.*, 309, 764, 1983.
16. **Fox, I. H. and Kelley, W. N.**, Studies on the mechanism of fructose-induced hyperuricemia in man, *Metabolism*, 21, 713, 1972.
17. **Mäenpää, P. H., Raivio, K. O., and Kekomäki, M. P.**, Liver adenine nucleotides: fructose-induced depletion and its effects on protein synthesis, *Science*, 161, 1253, 1968.
18. **Bode, J. C., Zelder, O., Rumpelt, H. J., and Wittkamp, U.**, Depletion of liver adenosine phosphates and metabolic effects of intravenous infusion of fructose or sorbitol in man and in the rat, *Eur. J. Clin. Invest.*, 3, 436, 1973.
19. **Morris, R. C., Jr., Brewer, E. D., and Broter, C.**, Evidence of a severe phosphate depletion dependent disturbance of cellular metabolism in patients with hereditary fructose intolerance (HFI), *Clin. Res.*, 78, 556, 1980.
20. **Kogut, M. D., Roe, T. F., Won, W., and Donnell, G. N.**, Fructose-induced hyperuricemia: observations in normal children and in patients with hereditary fructose intolerance and galactosemia, *Pediatr. Res.*, 9, 774, 1975.
21. **Perheentupa, J., Raivio, K. O., and Nikkilä, E. A.**, Hereditary fructose intolerance, *Acta Med. Scand.*, Suppl. 542, 65, 1972.
22. **Schmidt, E., Schmidt, F. W., and Wildhirt, E.**, Fermentaktivitats-Bestimmungen in der menschlicken Leber, 6 Mitteilung, *Klin. Wochenschr.*, 37, 1221, 1959.
23. **Pitkänen, E. and Perheentupa, J.**, Eine biochemische Untersuchung über zwei Fälle von Fructointo-leranz, *Ann. Paediatr. Fenn.*, 8, 236, 1962.
24. **Hers, H.-G.**, Augmentation de l'activité de la glucose 6-phosphatase dans l'intolérance au fructose, *Rev. Int. Hépatol.*, 12, 777, 1961/1962.
25. **Shapira, F., Shapira, G., and Dreyfus, J. C.**, La lésion enzymatique de la fructosurie bénigne, *Enzymol. Biol. Clin.*, 1, 170, 1961/1962.
26. **Nikkilä, E. A., Somersalo, A., Pitkänen, E., and Perheentupa, J.**, Hereditary fructose intolerance, an inborn deficiency of liver aldolase complex, *Metabolism*, 11, 727, 1962.
27. **Perheentupa, J. and Raivio, K.**, Fructose-induced hyperuricaemia, *Lancet*, 2, 528, 1967.
28. **Raivio, K. O., Kekomäki, M. P., and Mäenpää, P. H.**, Depletion of liver adenine nucleotides induced by D-fructose: dose-dependence and specificity of the fructose effect, *Biochem. Pharmacol.*, 18, 2615, 1968.

29. **Morris, R. C., Jr.**, An experimental renal acidification defect in patients with hereditary fructose intolerance. I. Its resemblance to renal tubular acidosis, *J. Clin. Invest.*, 47, 1389, 1968.

30. **Gentil, C., Colin, J., Valette, A. M., Alagille, D., and Lelong, M.**, Etude du métabolisme glucidique au cours de l'intolérance héréditaire au fructose. Essai d'interprétation de l'hypoglucosémie, *Rev. Fr. Etud. Clin. Biol.*, 9, 596, 1964.

31. **Sacrez, R., Juif, J.-G., Metais, P., Sofatzis, J., and Dourof, N.**, Un cas mortal d'intolérance héréditaire au fructose. Etude biochimique et enzymatique, *Pédiatrie*, 17, 875, 1962.

32. **Butenandt, I.**, Fructoseintoleranz, *Verh. Dtsch. Ges. Inn. Med.*, 70, 591, 1964.

33. **Kaplan, M., Straus, P., Brevart, P., Gorouben, J.-C., and Bedu-Saada, J.**, Etude clinique et biologique d'un cas d'intolérance au fructose chez un nourisson, *Sem. Hop.*, 39, 2733, 1963.

34. **Morris, R. C., Jr.**, An experimental renal acidification defect in patients with hereditary fructose intolerance. II. Its distinction from classical renal tubular acidosis; its resemblance to the renal acidification defect associated with the Fanconi syndrome of children with cystinosis, *J. Clin. Invest.*, 47, 1648, 1968.

35. **Manz, F. and Schärer, K.**, Long-term management of inherited renal tubular disorders, *Klin. Wochenschr.*, 60, 1115, 1982.

36. **Levin, B., Snodgrass, G. J. A. I., Oberholzer, V. G., Burgess, E. A., and Dobbs, R. H.**, Fructosemia: observations on seven cases, *Am. J. Med.*, 45, 826, 1968.

37. **Marthaler, T. M. and Froesch, E. R.**, Hereditary fructose intolerance. Dental status of eight patients, *Br. Dent. J.*, 123, 597, 1967.

38. **Cornblath, M., Rosenthal, I. M., Reisner, S. H., Wybregt, S. H., and Crane, R. K.**, Hereditary fructose intolerance, *N. Engl. J. Med.*, 269, 1271, 1963.

39. **Hamill, P. V. V., Drizo, T. A., Johnson, C. L., Reed, R. B., and Rock, A. F.**, NCHR growth charts: health examination survey data. Monthly Vital Statistics Report, Vol. 25 Supplement, Publ. No. (HRA) 76-1120, Public Health Service, Department of Health, Education and Welfare, Hyattsville, Md., 1976, 1.

40. **Raivio, K., Perheentupa, J., and Nikkilä, E. A.**, Aldolase activities in the liver in parents of patients with hereditary fructose intolerance, *Clin. Chim. Acta*, 17, 275, 1967.

41. **Steinmann, B. and Gitzelmann, R.**, The diagnosis of hereditary fructose intolerance, *Helv. Paediatr. Acta*, 36, 297, 1981.

42. **Borrone, C., Lamedica, G., Di Rocco, M., Conini, S., and Zanelli, C.**, Clinical heterogeneity in fructose intolerance, *Pediatr. Med. Chir.*, 4, 195, 1982.

43. **Andersson, G., Brohult, J., and Sterner, G.**, Increasing metabolic acidosis following fructose infusion in two children, *Acta Paediatr. Scand.*, 58, 301, 1969.

44. **Müller-Wiefel, D. E., Steinemann, B., Holm-Hadulla, M., Wille, L., Schärer, K., and Gitzelmann, R.**, Infusion associated kidney and liver failure in undiagnosed hereditary fructose intolerance, *Dtsch. Med. Wochenschr.*, 108, 985, 1983.

45. **Wagner, K. and Wolf, A. S.**, Death following fructose and sorbitol infusions, *Anaesthesist*, 33, 573, 1984.

46. **Odièvre, M., Gentil, C., Gautier, M., and Alagille, D.**, Hereditary fructose intolerance in childhood: diagnosis, management, and course in 55 patients, *Am. J. Dis. Child.*, 131, 605, 1978.

47. **Baerlocher, K., Gitzelmann, R., Steinmann, B., and Gitzelmann-Cumarasamy, N.**, Hereditary fructose intolerance in early childhood: a major diagnostic challenge: survey of 20 symptomatic cases, *Helv. Paediatr. Acta*, 33, 465, 1978.

48. **Lasker, M.**, Essential fructosuria, *Hum. Biol.*, 13, 51, 1941.

49. **Sachs, B., Sternfeld, L., and Kraus, G.**, Essential fructosuria: its pathophysiology, *Am. J. Dis. Child.*, 63, 252, 1942.

50. **Marble, A.**, Diagnosis of less common glycosurias, including pentosuria and fructosuria, *Med. Clin. N. Am.*, 31, 313, 1947.

51. **Silver, S. and Reiner, M.**, Essential fructosuria: report of 3 cases with metabolic studies, *Arch. Intern. Med.*, 54, 412, 1934.

52. **Gitzelmann, R., Steinmann, B., and Van den Berghe, G.**, Essential fructosuria, hereditary fructose intolerance, and fructose-1,6 diphosphate deficiency, in *The Metabolic Basis of Inherited Disease*, 5th ed., Stanbury, J. B., Wyngaarden, J. B., Frederickson, D. S., Goldstein, J. L., and Brown, M. S., Eds., McGraw-Hill, New York, 1983, 118.

53. **Heeres, P. A. and Vos, H.**, Fructosuria, *Arch. Intern. Med.*, 44, 47, 1929.

54. **Baylon, H., Shapira, F., Wegmann, R., Dreyfus, J. C., Moulias, R., Poyart, C., and Coumel, P.**, Note préliminaire sur l'étude clinique, biologique, histochimique et enzymatique de la fructosurie familiale essentielle, *Rev. Fr. Etud. Clin. Biol.*, 7, 531, 1962.

55. **Edhem, E. F., and Steinitz, K.**, Etudes sur un cas de levulosurie essentielle, *Acta Med. Scand.*, 97, 455, 1938.

56. **Baker, L. and Winegrad, A. I.**, Fasting hypoglycemia and metabolic acidosis associated with deficiency of hepatic fructose-1,6-phosphatase activity, *Lancet*, 2, 13, 1970.

57. **Corbeel, L. M., Eggermont, E., Bettens, W., Casteels-Van Daele, M., and Timmermans, J.,** Fructose intolerance with normal liver aldolase, *Helv. Paediatr. Acta*, 42, 626, 1970.

58. **Pagliara, A. S., Karl, I. E., Keating, J. P., Brown, B. I., and Kipnis, D. M.,** Hepatic fructose-1,6-diphosphatase deficiency, a cause of lactic acidosis and hypoglycemia in infancy, *J. Clin. Invest.*, 51, 2115, 1972.

59. **Berdanier, C. D.,** Genetic errors in carbohydrate metabolism, in *Carbohydrate Metabolism, Regulation and Physiological Role*, Berdanier, C. D., Ed., Hemisphere, Washington, D.C., 241, 1976.

60. **Enser, M., Shapiro, S., Horecker, B. L.,** Immunological studies of liver, kidney, and muscle fructose-1,6-diphosphatases, *Arch. Biochem. Biophys.*, 129, 377, 1969.

61. **Nishikawa, M., Tsukiyama, K., Enomoto, T., Tarui, S., Okuno, G., Ueda, K., Ikura, T., Tsujii, T., Sugase, T., Suda, M., and Tanaka, T.,** A new type of skeletal muscle glycogenosis due to phosphofructokinase deficiency, *Proc. Jpn. Acad.*, 41, 350, 1965.

62. **Tarui, S., Akuno, G., Ikura, Y., Tanaka, T., Suda, M., and Nishikawa, M.,** Phosphofructokinase deficiency in skeletal muscle. A new type of glycogenosis, *Biochem. Biophys. Res. Commun.*, 19, 517, 1965.

63. **Tarui, S., Kono, N., Nasu, T., and Nishikawa, M.,** Enzymatic basis for the coexistence of myopathy and hemolytic disease in inherited muscle phosphofructokinase deficiency, *Biochem. Biophys. Res. Commun.*, 34, 77, 1969.

64. **Layzer, R. B., Rowland, L. P., and Ranney, H. M.,** Muscle phosphofructokinase deficiency, *Arch. Neurol.*, 17, 512, 1967.

Chapter 5

GLUCOSE TOLERANCE

I. INTRODUCTION

Although diagnosis of clinical glucose intolerance and diabetes is based on elevations of fasting and postprandial blood glucose levels, blood insulin levels may be a more sensitive indicator of impairment of glucose tolerance. Elevated insulin levels in response to a glycemic stress have been associated with maturity onset (type II) diabetes and are considered to be one of the earliest detectable signs of diabetes.[1,2] Recent population studies in Finland,[3] Australia,[4] and France[5] have shown that hyperinsulinism is an independent risk factor for coronary heart disease. These epidemiological observations are in accord with experimental evidence showing that the arterial wall is an insulin-sensitive tissue and responds to high levels of insulin with the development of lipid-filled lesions.[6] The findings of increased incidence of coronary heart disease in subjects with impaired[7] and abnormal glucose tolerance in many patients with cardiovascular disease[8] suggest a relationship between these diseases which may be due to hyperinsulinism. Reduced insulin sensitivity has been postulated to be the primary step that leads to hyperinsulinemia.[9]

The effect of fructose on indices of glucose tolerance is of particular importance, since fructose is considered to be a sugar that can be consumed safely by diabetics. In this chapter the acute responses and the adaptive effects observed after the intake of fructose by experimental animals and humans on the levels of blood glucose and insulin and on insulin sensitivity will be described, and mechanisms proposed to explain these metabolic effects will be discussed.

II. EXPERIMENTAL ANIMALS

A. Feeding Studies

Table 1 summarizes the results from 17 studies in which the effects of dietary fructose on indices of glucose tolerance and histopathological changes associated with diabetes have been determined in the rat. In 7 of these studies no significant differences in the levels of blood glucose or insulin have been reported after rats were fed diets containing fructose as compared to glucose or starch.[10-16] However, a number of studies have demonstrated that the intake of fructose as compared to glucose or starch results in a deterioration of glucose tolerance[17-24] and histopathological changes associated with diabetes.[25,26] Affected parameters included higher levels of blood glucose and insulin both in the fasting state and after a glycemic stress. It is noteworthy that although fructose is metabolized primarily in the liver of the rat and its concentration in the systemic blood is low even after the intake of a large amount of fructose,[27] fructose feeding decreased the insulin sensitivity of the diaphragm and adipose tissue as compared to glucose feeding.[24]

The strongest experimental evidence that the nature of the dietary carbohydrate can be an etiological factor in diabetes comes from the work of Cohen et al.[28] using genetically selected rats. Rats were bred on the basis of blood glucose levels following an intragastric glucose tolerance test. Animals with the greatest increase in blood glucose (Upward selection) were mated, as were those with the smallest increase (Downward selection). In succeeding generations of the Upward selection fed a diet containing 72% fructose, the rats developed diabetes-like symptoms including postprandial hyperglycemia in excess of 250 mg % and diffuse glomerulosclerosis.[17] In contrast, comparable rats of the Upward selection fed a diet containing 72% cornstarch did not show these symptoms. In the offspring of the Downward

Table 1
EFFECT OF DIETARY FRUCTOSE ON INDICES OF GLUCOSE TOLERANCE AND HISTOPATHOLOGICAL CHANGES ASSOCIATED WITH DIABETES IN RATS

n	Dietary conditions	Results	Comments	Ref.
At least 7	72% fructose, glucose, or tapioca starch for 7 months	No significant differences in fasting blood glucose		10
8	68% fructose, glucose, or starch for 30 days	No significant differences in plasma insulin	Rats not fasted	11
6	70% of calories as fructose or glucose	No significant differences in serum insulin	Rats not fasted	12
10—12	68% fructose or glucose for 26 days	No significant differences in plasma glucose	Rats not fasted	13
6	68% fructose or glucose fed *ad libitum* or in meals for 36 days	No significant differences in fasting plasma glucose		14
8	66.5% fructose or glucose for 8—10 weeks	No significant differences in plasma insulin	Rats not fasted	15
16—43	66% fructose or glucose for 7 days	No significant differences in fasting plasma insulin or glucose		16
At least 12	72% fructose or cornstarch for 2 months	Glucose and insulin responses to a glucose or fructose load always significantly greater after feeding fructose	Genetically selected rats (Upward selected line)	17
At least 22	72% fructose or glucose for 2 months	Glucose responses 30 and 60 min after a glucose or fructose load significantly greater after fructose feeding; insulin responses 60 min after a glucose or fructose load significantly greater after feeding fructose	Genetically selected rats (Upward selected line)	18
6	70% of calories from fructose or glucose for 4 weeks	Glucose responses 30, 60, and 120 min after a glucose load significantly higher in rats adapted to fructose	Female rats	19
At least 9	66% fructose or glucose for 1 week	Fasting insulin significantly higher after fructose feeding; fasting glucose not significantly different		20
At least 9	66% fructose or 60% vegetable starch for 1 week	Insulin response to a glucose load significantly greater after fructose feeding, glucose response significantly higher only at 180 min; insulin and glucose responses to a fructose load significantly greater after fructose feeding; insulin sensitivity decreased after fructose feeding	Fructose compared to starch contained in rat chow	21
At least 8	15% fructose plus 39 or 54% cornstarch for times up to 15 months	Fasting insulin and glucose and the insulin response to a glucose load significantly higher after fructose feeding for 3, 5, 7, 9, and 15 months; glucose response to a glucose load significantly greater only at 7 months in fructose-fed rats		22
At least 23	66% fructose or 60% vegetable starch for at least 1 week	Glucose and insulin levels significantly higher in fructose-fed rats $4\frac{1}{2}$ hr after removal of food	Fructose compared to starch in rat chow	23

Table 1 (continued)
EFFECT OF DIETARY FRUCTOSE ON INDICES OF GLUCOSE TOLERANCE AND HISTOPATHOLOGICAL CHANGES ASSOCIATED WITH DIABETES IN RATS

n	Dietary conditions	Results	Comments	Ref.
6—7	70% of calories from fructose or glucose for 4 weeks	Insulin sensitivity of diaphragm muscle and parametrial adipose tissue significantly lower after rats consumed the fructose		24
6	68% fructose, starch, glucose, or fructose for 6 months	Retinopathy produced by fructose and sucrose, similar to that produced in diabetic rats; retinopathy not produced by starch and glucose		25
6	68% fructose, starch, glucose, or sucrose for 6 months	Incidence of diffuse renal glomerulosclerosis, tubular damage, and lymphocytic infiltration greater in rats fed fructose and sucrose than in those fed starch or glucose		26

selection the diabetes-like symptoms did not appear in rats fed either the 72% fructose or starch diets. A similar pattern, but of lesser magnitude, was observed when the rats were fed 72% fructose as compared to 72% glucose.[18] The results show that in these rats the interaction between genetic and dietary factors is necessary for the expression of the metabolic defects characteristic of diabetes, and suggest that similar interactions might influence expression of such defects in humans.

In their initial studies, Cohen et al.[28] found that rats of the Upward selection fed a diet containing 72% sucrose as compared to starch developed diabetic-like symptoms similar to those subsequently shown in fructose-fed rats. These findings imply that it is the fructose moiety of sucrose that is responsible for the impairment of glucose tolerance reported in studies in which rats are fed sucrose as compared to glucose-based carbohydrates.[29-34] This implication is supported by the findings of Boot-Hanford and Heath that fructose is the agent responsible for the retinopathy[25] and glomerulosclerosis[26] observed after feeding diets containing 68% sucrose to rats for 6 months. It is noteworthy that the addition of a diet rich in wheat and oats to the fructose diet prevented the occurrence of the glomerulosclerosis.[26] The authors partially attributed the protective effect to the presence of additional fiber. This is an example of interactions that can occur between nutrients which could explain apparently contradictory findings in studies using essentially the same dietary conditions.

The relevancy of the studies listed in Table 1 to human nutrition may be questioned, since in most cases fructose was fed at levels very much higher than normally consumed. The premise that high levels of any nutrient will produce adverse metabolic effects and that these adverse effects do not reflect a true dietary hazard is not necessarily true. The studies reported were comparative; a large amount of fructose was compared to the effects of an equally large amount of a glucose-based carbohydrate. Moreover, fructose feeding at levels as low as 15% of total calories for extended time periods was shown to increase both the insulin and glucose responses to a glucose load.[22] While the studies reported with rats are not necessarily pertinent to humans, previously observed similarities in carbohydrate metabolism between rats and humans warrant further studies with humans on the long-term effect of fructose on glucose tolerance before fructose can be recommended as a desirable source of dietary carbohydrate.

B. Possible Mechanisms of Fructose Action

The mechanisms by which dietary fructose may produce the reported deterioration of glucose tolerance in experimental animals are based on the observed effects of fructose on carbohydrate, lipid, and nucleotide metabolism. These studies have necessarily employed in vitro techniques such as the use of isolated tissues, but also have utilized in vivo response studies. In most of these studies fructose levels much in excess of those expected to be found in the systemic circulation following oral ingestion of fructose were used. Whether results obtained under these conditions are relevant to those that would be found after the feeding of lesser levels of fructose over extended time periods must be considered. The liver has necessarily been the focus for many of these studies, since the metabolism of fructose occurs primarily in this tissue. These studies have identified a number of metabolic effects of fructose that can result in elevated levels of glucose and insulin and reduced tissue sensitivity to insulin. Since these metabolic defects are interrelated and often concurrently present, it is difficult to assess which may be the primary metabolic defect. For example, the production and secretion of high levels of glucose by the liver can provoke the necessary insulin response from the pancreas which, if high enough, can initiate the down-regulation of insulin receptors[35] and resultant insulin insensitivity. Conversely, insulin insensitivity of tissues could initiate metabolic defects leading to glucose intolerance.[9]

Metabolic effects of fructose as they pertain to glucose tolerance can be divided into three general categories: (1) direct effects on insulin secretion, (2) effects through actions on carbohydrate metabolism, and (3) effects through actions on lipid metabolism. Underlying the metabolic effects of fructose are differences in genetic predisposition of animal models to cope with these effects.[17,18,36]

1. Direct Effects on Insulin Secretion

The ability of fructose to stimulate the secretion of insulin from the pancreas appears to depend on the presence of threshold levels of glucose. In the absence of glucose in the perfusing media, 300 mg/100 mℓ of fructose did not potentiate insulin release from isolated rat pancreas.[37] When glucose was added to the media at concentrations simulating fasting conditions (50 to 100 mg/100 mℓ), there was still no fructose-induced stimulation of insulin secretion. However, when glucose was infused to produce blood levels of 150 mg%, an amount that could occur in the postprandial state or in diabetics, the subsequent infusion of fructose produced a potentiation of the glucose-induced insulin release.[37] This finding reconciled apparently conflicting findings pertaining to the stimulatory effect of fructose on insulin secretion in a number of animal models.[37-39]

It is therefore theoretically possible to explain the relative hyperinsulinemia observed in fructose-fed rats if the following conditions are present: (1) blood glucose is elevated and (2) the necessary systemic blood fructose level is achieved. The necessary glucose concentrations can be attained either acutely under postprandial conditions or chronically by fructose-induced disturbances of glucose metabolism (see Section II.B.2) and/or decreased insulin sensitivity.[24] The minimum level of systemic fructose required to induce the potentiation of insulin secretion has yet to be determined. However, for this mechanism to explain fructose-induced hyperinsulinemia, this level must be much lower than that used in perfusion studies. Systemic blood concentrations of 2 to 6 mg/100 mℓ were reported 40 to 50 min following the gastric intubation of 2 g of fructose into rats.[27] However, systemic blood fructose levels as high as 27 mg/100 mℓ were found 60 to 90 min after a total of 6 g of fructose was intubated into rats during a 2-hr period.[40]

Another mechanism by which dietary fructose may produce a relative hyperinsulinemia is by being the precursor of a metabolite relatively specific for fructose metabolism and which has the capacity for stimulating insulin secretion. The hepatic metabolism of fructose, but not glucose, results in the formation of D-glyceraldehyde. Since this metabolite is not ionically charged, it might be expected to gain access to the systemic circulation. At levels as low as 2 to 4 mM, D-glyceraldehyde was a more potent secretagogue for the stimulation of insulin secretion from isolated rat pancreatic islets than was glucose.[41] Similar findings have been reported using mouse pancreas.[42,43]

2. Effects Through Actions on Carbohydrate Metabolism

A fructose load given to normoglycemic rats unadapted to fructose feeding does not produce a significant increase in blood glucose.[27,40] However, fructose feeding appears to produce changes in carbohydrate metabolism that favor the increased production of glucose and its decreased clearance from the circulation. These changes can provide the blood glucose levels required to provoke a hyperinsulinemic response either directly or by the synergistic action with fructose described above, even when glucose-based carbohydrates are not present in the diet.[17-20,24]

Since the metabolism of fructose occurs primarily in the liver, this tissue would be expected to be the site of the major effects of fructose on glucose metabolism. The liver appears to play a major role in the total utilization of glucose provided from dietary sources.[44] The incorporation of glucose into glycogen both in vitro[45] and in vivo[19,46] was shown to be significantly lower after rats had been fed a fructose than either a glucose or starch diet. Since a large portion of glucose derived from dietary sources is incorporated into liver glycogen,[47] an inhibition of this pathway for glucose utilization could produce elevated levels of blood glucose. The reduced incorporation of glucose into glycogen may be due to the decreased activity of glucokinase in rats fed fructose[48,49] or fructose-containing sugars[50] as compared to glucose-based carbohydrates. Glucokinase phosphorylates glucose to glucose-6-phosphate, thereby trapping it intracellularly. Glucose-6-phosphate is a precursor in the conversion of glucose to both glycogen and lipid. The decreased activity of glucokinase may be aggravated by the concurrent enhancement of the activity of fructokinase observed after the feeding of fructose[51,52] and the resultant decrease in hepatic ATP required for the hepatic phosphorylation of circulating glucose. The reduced hepatic glycogen synthesis and glucose oxidation observed after rats are fed fructose- as compared to glucose-based carbohydrates[19,53] is consistent with these metabolic events.

The activity of glucose-6-phosphatase, an important gluconeogenic enzyme, is increased when rats are fed fructose[54] or sucrose[55] as compared to glucose-based carbohydrates. The increased activity of this enzyme would favor hydrolysis of glucose-6-phosphate formed by glucokinase and favor the efflux of glucose from the liver to the blood and thus contribute to the overall decrease in hepatic glucose uptake. The activity of another key hepatic gluconeogenic enzyme, phosphoenolpyruvate carboxykinase, has been reported to be significantly increased after rats were fed fructose as compared to cornstarch for 3 months.[22] These enzymatic effects are consistent with the observed increased output of glucose from the liver of rats fed fructose as compared to starch.[23]

An important source of blood glucose is the breakdown of endogenous liver glycogen. The reported effects of fructose on liver glycogen are primarily based on in vitro infusion studies utilizing levels of fructose which would be unlikely to occur under dietary conditions. These studies have produced contradictory results. Fructose has been reported to increase the net rate of glycogen formation based primarily on an inhibition of the active form of phosphorylase, phosphorylase a, by high levels of fructose-1-phosphate.[56-58] In contrast, the in vitro perfusion[59,60] or in vivo infusion[61] of fructose into rats was reported to increase phosphorylase a activity and decrease the active form of glycogen synthase, thus promoting glycogen breakdown. The increase in phosphorylase a activity by fructose has been proposed to be due to a reduction of ATP,[62] producing a decrease in the ATP/Mg^{++} ratio.[59] It has been suggested that fructose will promote glycogenolysis when glycogen stores are full, but will increase glycogen stores when they are suboptimal.[63] In this regard it has been reported that rats given 4 g of fructose per kilogram body weight by gavage after a 24-hr fast show rapid increases in glycogen synthase activity and liver glycogen levels.[64] The relevance of these findings to the chronic dietary intake of fructose over prolonged time periods has yet to be determined.

It has been proposed that an increase in retinal lactate concentration is a causal agent in the development of diabetic retinopathy.[65] An increase in retinal lactate concentration and the lactate/pyruvate ratio has been observed in normal rats fed sucrose as compared to starch.[66] Since there were no significant correlations between the glucose and lactate content of the blood and retina, it was concluded that the increased retinal lactate in the sucrose-fed rats was derived from the fructose moiety of sucrose. In streptozotocin diabetic rats, the retinal concentration of fructose was significantly higher after the feeding of a diet containing 68% sucrose as compared to cornstarch.[67] Retinopathy due to an effect of fructose on vitamin A metabolism[68,69] does not appear likely, since both plasma and tissue levels of this vitamin have been reported to be normal in rats fed diets containing high levels of sucrose or fructose.[70]

3. Effects Through Actions on Lipid Metabolism

Muscle and adipose tissue also contribute significantly to the total utilization of orally administered glucose in the rat.[44] In these tissues, fructokinase is not present in significant amounts and the metabolism of fructose is initiated by phosphorylation at the 6-hydroxy position by hexokinase. Since the affinity of hexokinase for glucose is several orders of magnitude greater than for fructose,[71] physiological levels of glucose would be expected to suppress fructose phosphorylation. Adipose tissue may be an exception because of the very low concentration of free glucose in this tissue.[72] Despite the apparently poor affinity of hexokinase for fructose and the low concentration of fructose in the systemic circulation of fructose-fed rats,[27] the feeding of fructose as compared to glucose has been reported to decrease glucose utilization by adipose tissue and, to a lesser extent, muscle in rats.[24]

Since the metabolism of glucose by adipose tissue and muscle is insulin-dependent, the decreased utilization of glucose in these tissues from fructose-fed rats may be the result of a decreased insulin sensitivity. In this regard the insulin stimulation of the uptake and metabolism of glucose by adipose tissue[24,73] and diaphragm[24] was significantly lower after rats consumed diets containing fructose as compared to glucose. It has been reported that hypertriglyceridemic subjects bound 34% less insulin to monocytes than did normotriglyceridemic subjects of comparable age and body weight.[74] Fructose may therefore decrease insulin sensitivity of adipose and muscle tissue indirectly through its well-established ability to increase hepatic lipogenesis and the levels of blood triglycerides (see Chapter 7).

It has been shown that the release of free fatty acids from adipose tissue of rats fed fructose is significantly greater than of those fed glucose.[12] The enhanced release of the free fatty acids was associated with their increased concentration in the blood of the fructose-fed rats. Adipose tissue lipolysis and reesterification are influenced by insulin, and the increased mobilization of free fatty acids may be a reflection of the reduced insulin sensitivity of fructose-fed rats. Alternatively, the increased production of free fatty acids may be in response to the need for alternative metabolic fuels in insulin-sensitive tissues (e.g., muscle) which usually utilize glucose. Increased levels of circulating free fatty acids have been shown to inhibit glucose utilization,[75] and would therefore be expected to further exacerbate the insulin insensitivity of fructose-fed rats.

III. HUMAN STUDIES

A. Response Studies

Many reports of the effects of fructose on indices of glucose tolerance in humans have been based on responses observed after the acute ingestion of a fructose load. This load is usually given after an overnight fast, and the effects on blood levels of glucose, insulin, and fructose are compared to those observed after an equivalent load of another carbohydrate

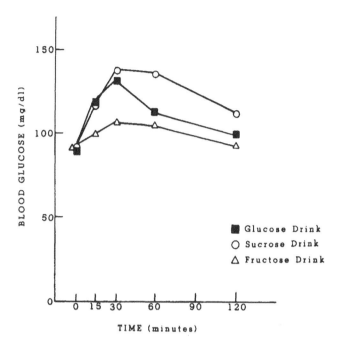

FIGURE 1.   Plasma glucose responses of nine subjects after consuming a drink containing 100 g of glucose, sucrose, or fructose. (Adapted from Bohannon, N. V., Karam, J. H., and Forsham, P. H., *J. Am. Diet. Assoc.*, 76, 555, 1980.)

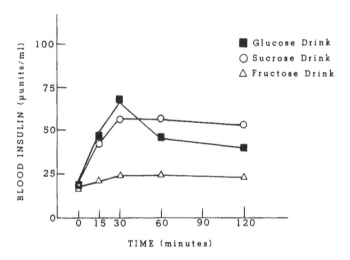

FIGURE 2.   Serum insulin responses of nine subjects after consuming a drink containing 100 g of glucose, sucrose, or fructose. (Adapted from Bohannon, N. V., Karam, J. H., and Forsham, P. H., *J. Am. Diet. Assoc.*, 76, 555, 1980.)

is given to the same subjects under the same experimental conditions. Under these conditions fructose elicits a much lower response of blood glucose and insulin than does a carbohydrate containing glucose, in both normal subjects[76-79] and those exhibiting signs of glucose intolerance.[79,80] Figure 1 presents the blood glucose responses typical of those reported after subjects consumed a drink containing 100 g of glucose, sucrose, or fructose. Figure 2 shows the corresponding blood insulin responses. As with the rat, systemic blood fructose levels

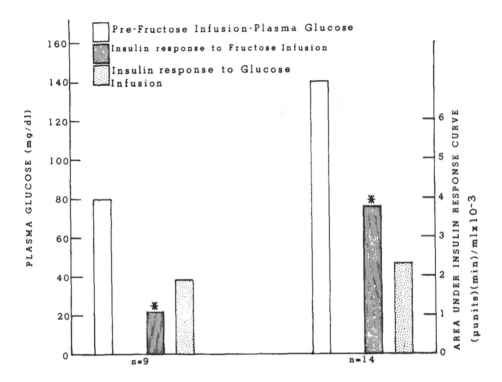

FIGURE 3.   Effect of fructose infusion on insulin output (area under response curve) in normal subjects as a function of blood glucose concentration. (*) Significantly different from insulin response to glucose infusion ($p < 0.01$). (Adapted from Lawrence, J. R., Gray, C. E., Grant, I. S., Ford, J. A., McIntosh, W. B., and Dunnigan, M. G., *Diabetes*, 29, 736, 1980.)

remain low even after the oral intake of large amounts of fructose. Maximal levels of systemic blood fructose have been reported to be between 10 and 12 mg % 1 hr after the intake of 1 g fructose per kilogram body weight.[76,78,81] The excretion of urinary glucose by subjects with impaired glucose tolerance or noninsulin-dependent diabetes was lower after they consumed a fructose drink than either a sucrose or glucose drink.[79] These response studies have been the basis for the conclusion that including fructose in the diet of diabetics would have beneficial effects. It might also follow that the consumption of fructose instead of glucose-based sugars would be advantageous to normal individuals in a population such as ours in the U.S. in which the incidence of diabetes increases with age.

Since fructose is rarely consumed alone, the metabolic effects of fructose in the presence of other food components are more relevant to actual dietary conditions. Fructose, whether incorporated into a meal as either sucrose or high-fructose corn sweeteners, is usually consumed concurrently with glucose-yielding carbohydrates which would be expected to produce postprandial elevations of blood glucose. Using i.v. infusion techniques, it has been shown that the ability of fructose to provoke insulin secretion in humans is augmented when blood glucose concentrations are elevated.[38,39,82,83] This is best illustrated by the study of Lawrence et al.[39] (Figure 3). When prefructose-infusion blood glucose was maintained in normal subjects at levels approximating fasting, the subsequent infusion of 30 g of glucose as compared to the same amount of fructose produced a nearly twofold increase in insulin output. In contrast, when glucose was infused to simulate postprandial blood glucose levels, the insulin output following infusion of fructose was significantly greater than that following infusion of glucose. In these subjects the blood insulin levels 10 to 50 min after the fructose infusion showed a significant correlation with blood fructose levels ($r = 0.93$; $p < 0.01$) but an insignificant correlation with blood glucose levels ($r = 0.10$). Blood glucose levels

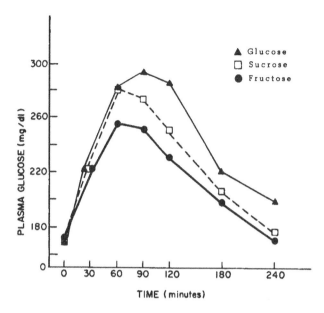

FIGURE 4.   Mean plasma concentrations of glucose before and after ten noninsulin-dependent diabetics consumed the sugar-containing meals. (Adapted from Bantle, J. P., Laine, D. C., Castle, G. W., Thomas, J. W., Hoogwerf, B. J., and Goetz, F. C., *N. Engl. J. Med.*, 309, 7, 1983.)

during fructose infusion were not greater than during glucose infusion and did not rise at any time after fructose infusion to glucose-primed subjects. A similar pattern of blood glucose-dependent, fructose-induced stimulation of insulin output was also observed in noninsulin-dependent diabetic subjects.[39] In evaluating the nutritional significance of these infusion studies the same considerations apply as discussed above for the analogous situation in the rat.[37] While blood glucose levels approximating those in Figure 3 can be obtained postprandially, the systemic fructose concentrations were as much as tenfold greater following fructose infusion than those reported after fructose ingestion. Whether the chronic feeding of diets containing high levels of fructose can produce the systemic fructose concentrations required to simulate the conditions attained by infusion and thereby significantly increase insulin levels remains to be determined.

A number of studies have been performed in which the responses to fructose, as compared to other carbohydrates incorporated in a meal, on indices of glucose tolerance have been determined.[79,80,84-88] The results of these studies are often difficult to interpret because of the inclusion of different food components in the meals; however, they suggest that the insulin response to fructose incorporated in a meal is much closer to that of other carbohydrates than when given alone. For example, Bantle et al.[87] fed both diabetics and normal subjects a test meal containing about 25% of the total calories as glucose, sucrose, fructose, or starch. The specific foods contained in the meal included two eggs, 30 g of Canadian bacon, 10 g of margarine, 210 g of cooked cream of rice containing 30 g of rice starch and 240 mℓ of skim milk, and 180 mℓ of black coffee. The sugar-containing meals had a total of 685 kcal — 49% from carbohydrate, 33% from fat, and 18% from protein. Figures 4 and 5 present the glucose and insulin responses, respectively, to the sugar-containing test meals of ten noninsulin-dependent diabetics. As compared to the glucose meal and, to a lesser extent, the sucrose meal, the fructose meal produced the smallest peak blood glucose values and the smallest area under the glucose curve (Figure 4). In contrast, the peak insulin levels after the fructose meal were not significantly different from those obtained after the

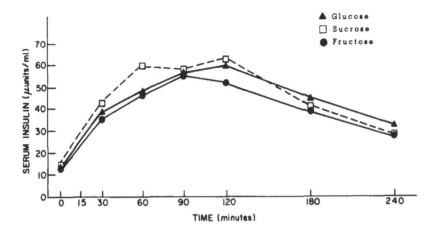

FIGURE 5.   Mean serum concentrations of insulin before and after ten noninsulin-dependent diabetics consumed the sugar-containing meals. (Adapted from Bantle, J. P., Laine, D. C., Castle, G. W., Thomas, J. W., Hoogwerf, B. J., and Goetz, F. C., *N. Engl. J. Med.*, 309, 7, 1983.)

glucose or sucrose meal, and the areas under the insulin curve appeared similar (Figure 5). The glucose responses of ten healthy subjects to the sugar-containing meals were similar to those obtained with the diabetics. Although there were no significant differences in peak insulin levels after the sugar-containing meals, normal subjects showed an apparently greater area under the insulin curve after consuming the glucose meal than after consuming the fructose meal. These results indicate that diabetics and, to a lesser extent, healthy subjects respond to a fructose meal with insulin responses that appear to be inappropriate for the concurrent increase in blood glucose levels. Since fructose appears to be a poor secretagogue for gastric inhibitory polypeptide (GIP),[63,89] a fructose-induced effect on insulin secretion via the enteroinsular axis appears unlikely. Since blood glucose levels in the diabetics reached levels as high as 250 mg % after the fructose meal, conditions appeared to be conducive for the synergistic effect of fructose on insulin secretion noted in the infusion studies described above.

In a series of studies by Akgün and Ertel,[84,85,88] glucose and insulin responses were determined after both normal and diabetic subjects consumed meals containing sucrose, high-fructose corn syrup, or fructose. The meals contained about 400 kcal and were composed of scrambled egg, farina, low-fat milk, decaffeinated coffee, and 35 g equivalent of the sugars. In noninsulin-dependent diabetics, the meals containing sucrose[84,88] or high-fructose corn sweeteners[85,88] gave significantly greater areas under the glucose response curves than did the meal containing fructose. In contrast, there were usually no significant differences in the corresponding areas under the insulin response curves after the diabetics consumed the sugar-containing meals. These results are in general accord with those of Bantle et al.[87] In normal subjects,[84] the increments in both plasma insulin and glucose were greater after the sucrose- than the fructose-containing meal, emphasizing the importance of the level of blood glucose attained after a meal as a determinant in the ability of fructose to promote insulin secretion.

In a series of studies by Crapo and colleagues, the comparative effects of meals containing fructose, as compared to either glucose[79] or sucrose,[79,80,86] on blood glucose and insulin responses were determined. After a meal containing 50 g of sugar, 20 g of corn oil, and 20 g of egg albumin, the glucose and insulin responses following the fructose meal were significantly lower than those observed following either the glucose or sucrose meal in both normal subjects and those with impaired glucose tolerance.[79] The responses after the fructose meal and a fructose drink were also similar in these subjects. The failure of the fructose meal to provoke an increased insulin response may be due to the absence of a dietary component capable of raising blood glucose to the necessary levels. With noninsulin-dependent diabetics, meals containing glucose and sucrose produced significantly greater glucose responses than did the meal containing fructose. In contrast, there were no significant differences in insulin responses among the three sugars in the diabetics; however, insulin responses in these subjects were extremely low. A pattern of insulin and glucose responses similar to those found in the previous study[79] was observed in normal subjects, subjects with impaired glucose tolerance, and noninsulin-dependent diabetics after they consumed cake or ice cream containing either sucrose or fructose.[86] The cakes contained 63 g of sugar, 28 g of other carbohydrates (predominantly starch), 21 g of fat, and 7 g of protein. The ice cream was composed of 52 g of either sucrose or fructose, 10 g of lactose, 24 g of fat, and 10 g of protein. Feeding cakes similar in composition to those described above to subjects with reactive hypoglycemia produced significantly reduced glucose and insulin responses after the fructose as compared to the sucrose cake.[80]

The results obtained from studies of the effects of a fructose-containing meal on indices of glucose tolerance in both normal and glucose-intolerant subjects emphasize that the ability of fructose to increase insulin levels is dependent on blood glucose concentrations. In view of the undesirable effects of high levels of insulin, it would appear that any beneficial effects of fructose on blood glucose levels in glucose-intolerant subjects and those with noninsulin-dependent diabetes should be evaluated with regard to concurrent effects on blood insulin levels.

B. Feeding Studies

There have been surprisingly few studies in which the metabolic effects of the extended feeding of fructose incorporated into a diet similar in composition to that consumed by humans have been determined. These types of studies are important in evaluating the desirability of fructose consumption in view of the adaptive effects on enzymatic and physiological processes evoked by chronic fructose feeding and in view of the interactions between fructose and other dietary components that may occur and thereby influence metabolic processes. These human studies generally can be divided into two categories. The first involves the comparative effects of fructose and sucrose feeding on indices of glucose tolerance; these studies are based on the practical consideration of which sweetener would provide more desirable metabolic effects. The other category involves comparisons between fructose and glucose-based carbohydrates; these studies are based mainly on metabolic considerations, especially in view of the large differences in glycemic responses exhibited by sugars and starches from various food sources.[90] The results of these studies will be described in detail in this section.

1. Comparisons with Sucrose

A 2-year study on the comparative effects of fructose, sucrose, and xylitol on indices of glucose tolerance was conducted in Finland.[91] Volunteers from 13 to 55 years of age were divided into three groups and instructed to use fructose ($n = 35$), sucrose ($n = 33$), or xylitol ($n = 48$) as the only sweetening agent in their otherwise self-selected diets. The average daily intake of the sweeteners ranged from 59 to 73 g for sucrose, 56 to 69 g for fructose, and 43 to 50 g for xylitol, with the lower levels observed during the final 8 months of the study. Fasting levels of blood glucose and insulin were measured after the 11th, 14th, 18th, and 22nd months of the study. Since the initial fasting levels for glucose and insulin for the three groups were not reported, it is difficult to evaluate these results across diet. However, it was reported that the fasting blood glucose concentrations were similar in the dietary groups and that no significant differences were found in blood insulin. Likewise, a glucose tolerance test (1 g/kg body weight) given to subjects after consuming their respective

Table 2

GLUCOSE RESPONSES TO A GLUCOSE TOLERANCE TEST (1 g GLUCOSE PER KILOGRAM BODY WEIGHT) BY SUBJECTS CONSUMING FRUCTOSE, SUCROSE, OR XYLITOL FOR 14 MONTHS

Sweetener	Blood glucose responses (mg %)		
	0 hr	1 hr	2 hr
Fructose (n = 35)	76 ± 1.8	97 ± 5.4	82 ± 3.6
Sucrose (n = 33)	77 ± 1.8	88 ± 5.4	82 ± 1.8
Xylitol (n = 48)	81 ± 1.8	94 ± 3.6	81 ± 3.6

Note: Mean ± SEM from number of subjects shown in parentheses.

Adapted from Huttunen, J. K., Mäkinen, K. K., and Scheinin, A., *Acta Odont. Scand.*, 33 (Suppl. 70), 239, 1975.

sweeteners for 14 months resulted in no significant differences in glucose (Table 2) and insulin responses. A number of experimental defects prevent definitive interpretation of the results of this study as proving that the long-term feeding of sucrose and fructose produces similar effects on glucose tolerance. The experimental design did not include a crossover which would have enabled a comparison of the effects of the sweeteners after they were consumed by each of the subjects. Even at maximal intake, the sweeteners were present in only 10% of the calories. The different dietary components in the self-selected diets consumed by the subjects could have affected the action of the sweeteners on glucose tolerance. There also appeared to be poor control of the sweetener intake by the subjects.

In a study reported by Bossetti et al.,[92] eight subjects (four male, four female) 22 to 32 years of age and free of overt disease were fed either fructose or sucrose incorporated into a diet comprised of typical American foods for 14 days each in a crossover design. The prestudy diets of the subjects were analyzed to determine the composition of the diet used for each subject during the study. The sugars comprised 13% of the 35 to 49% of the calories provided by carbohydrate. The levels of the sugars ranged from 50 to 107 g/day. Other components of the diet included fat (35 to 45% of calories), protein (12 to 20% of calories), and fiber (less than 10 g/day). The P/S ratio of the diets ranged from 0.5 to 1.5. Fasting levels of blood glucose and insulin were determined on days 7 and 14 of each dietary period. There were no significant differences in glucose, insulin, or the insulin-to-glucose ratio after the subjects consumed either sugar. From these results, the investigators concluded that they would not suggest the preferential use of either sugar in healthy individuals.

In a study using a somewhat different experimental protocol,[93] 11 subjects (4 male, 7 female) 29 to 62 years of age were fed a diet containing 13% of the calories as sucrose for 3 to 4 days (base line period), after which the sucrose was replaced by fructose and the diets fed for 14 more days. During the study, each subject consumed a weight-maintaining diet that contained 55% of calories from carbohydrate, 30% from fat, and 15% from protein. The diets provided 300 mg of cholesterol and 25 g of dietary fiber daily and had a P/S ratio of about 1. The daily range of sucrose or fructose provided by the diets was 63 to 99 g. During the base line period and after 3 and 14 days of consuming the fructose-containing diet, glucose and insulin responses were determined during oral glucose and fructose tolerance tests and after the breakfast and lunch meals. Following the glucose tolerance test, serum glucose values after 14 days of fructose feeding were not significantly different from base

Table 3
EFFECT OF DIETARY STARCH AND FRUCTOSE ON FASTING BLOOD GLUCOSE AND URINARY GLUCOSE EXCRETION IN UNTREATED AND INSULIN-TREATED ADULT DIABETICS

Diabetics	Glucose parameter	Period	
		Starch	Fructose
Untreated ($n = 6$)	Fasting (mg %)	203	218
Treated ($n = 10$)	Fasting (mg %)	140	155
	Urinary (g/day)	15.1	15.7

Note: Each value represents the mean from the number of diabetics shown in parenthesis.

Adapted from Nikkilä, E. A., in *Sugars in Nutrition,* Sipple, H. L. and McNutt, K. W., Eds., Academic Press, Orlando, Fla., 1974, chap. 26.

line values. The corresponding insulin responses were significantly lower at 180 min after fructose feeding only. Serum glucose was significantly lower at fasting and 30 min following the fructose tolerance test after the subjects had consumed the fructose diet for 14 days than during the base line period; however, insulin levels were not different. The only significant difference noted in glucose and insulin responses following the meals was 1 hr after lunch, when both parameters were significantly lower after fructose feeding for 14 days then they were during the base line period. It would have been of interest to compare the effect of these sugars after equivalent feeding periods or after using a nonfructose-containing carbohydrate during the base line period, since there was an increase of only 6.5% in the caloric content of fructose in the experimental as compared to the base line diet.

These studies have not demonstrated a particular metabolic advantage for the use of fructose over sucrose as a sweetener, especially with respect to insulin responses. Although studies of the effect of the extended feeding of sucrose as compared to glucose or glucose polymers on insulin levels have not given conclusive results, 8 out of 12 reported an increase in insulin due to sucrose.[94] If, as indicated from these studies, sucrose and fructose produce equivalent effects on insulin status, then it would appear permature to advocate the use of fructose as a sweetener until more definitive studies are performed.

2. Comparisons with Glucose-Based Carbohydrates

Nikkilä[95] reported on the effects of feeding starch as compared to fructose on fasting blood glucose and urinary glucose excretion in untreated and insulin-treated, adult-onset diabetics. During the starch period, the diet contained 45% of calories as carbohydrate (80% starch), 35% as fat (mainly saturated), and 20% as protein. During the fructose period the diabetics were fed the same basic diet but with 75 to 90 g of starch replaced by fructose. Fructose was added to the diet as a powder on desserts or in beverages. Each dietary period was for 10 to 20 days. The results are shown in Table 3. Although there was a tendency for higher blood glucose levels during the fructose as compared to the starch period, these increases were not statistically significant. On the basis of these results, it was concluded that diabetics can use a moderate amount of fructose in their diet without any serious short-term undesirable metabolic effects. However, it was pointed out that the metabolic effects of the fructose varied markedly according to individual subjects, with some diabetics showing a marked increase of both blood glucose levels and urinary glucose excretion while others were

Table 4

**BODY WEIGHT, FASTING BLOOD GLUCOSE, INSULIN, INSULIN
BINDING TO MONOCYTES, AND INSULIN SENSITIVITY BEFORE AND
7 DAYS AFTER SUBJECTS CONSUMED SELF-SELECTED DIETS PLUS
AN EXTRA 1000 KCAL AS EITHER GLUCOSE OR FRUCTOSE**

Parameter	Fed excess glucose ($n = 7$)		Fed excess fructose ($n = 8$)	
	Base line	After diet	Base line	After diet
Body weight (kg)	60.9	61.4	61.5	62.0
Fasting plasma glucose (mg %)	104	97	94	92
Fasting plasma insulin (μU/mℓ)	7	7	4	3
Insulin binding to monocytes[a]	3.3	3.0	3.2	2.0
				$p < 0.05$
Insulin sensitivity[b]	5.5	4.9	6.0	4.6
				$p < 0.05$

Note: Each value represents the average from the number of subjects shown in parentheses.

[a]   Expressed as specific cellbound fraction $\times 10^{-2}$.
[b]   Expressed as the rate constant for glucose disappearance after i.v. injection of 0.05 U of insulin per kilogram body weight.

Adapted from Beck-Nielsen, H., Pedersen, O., and Lindskov, H. O., *Am. J. Clin. Nutr.*, 33, 273, 1980.

completely unaffected by the sugar. It would appear difficult, therefore, to make any blanket recommendation to diabetics or any subject group concerning fructose intake that would be equally relevant to each individual in that group.

In a 4-day study of 28 patients with diagnosed coronary artery disease fed a diet containing approximately 62% of the calories as carbohydrate, 22% as fat, and 16% as protein and providing less than 200 mg of cholesterol daily, the inclusion of 2 g of glucose per kilogram body weight ($n = 14$) did not significantly affect fasting blood glucose as compared to pretest levels (94 mg % on each diet).[96] In contrast, the inclusion of 2 g of fructose per kilogram body weight ($n = 14$) produced a small but significant reduction in fasting blood glucose from pretest levels of 94 to 90 mg %. The physiological significance of this decrease in blood glucose in these subjects should be determined after the feeding of fructose for more extended periods of time in the context of a diet more comparable to that presently consumed.

In a study to determine the effects of various dietary components on indices of glucose tolerance, it was found that the addition of 1000 kcal as sucrose to the self-selected diets of six normal subjects for 14 days significantly decreased both insulin binding to isolated monocytes and insulin sensitivity.[97] The subjects were reported to have gained an average of 1.7 kg during this 2-week period. A follow-up study was then carried out, aimed at determining which of the monosaccharide moieties of sucrose was responsible for these effects on insulin binding and sensitivity.[98] Young (21 to 35 years of age), healthy subjects (mainly females) within 80 to 120% of their ideal weight were randomly divided into two groups and instructed to consume either 1000 kcal (250 g) of glucose ($n = 7$) or an equivalent amount of fructose (n = 8) in addition to their self-selected diets for 7 days. The sugars were given four times daily dissolved in water. The diets consumed by the subjects during the study provided an average of 44% of calories from carbohydrate, 38% from fat, and 18% from protein. Effects on body weight, fasting blood glucose and insulin, insulin binding to isolated monocytes, and insulin sensitivity expressed as the rate constant for plasma glucose disappearance in response to intravenously injected insulin were determined before and after the 7-day dietary period (Table 4). Before the study began there were no statistically significant differences in the parameters measured between the subjects to be fed glucose or fructose. The feeding of excess glucose produced no significant changes in any of the parameters measured relative to base line levels. The feeding of excess fructose produced no significant differences in body weight and fasting levels of blood glucose and insulin as compared to base line values. However, fructose feeding was accompanied by significant decreases of both insulin binding and insulin sensitivity. The differential effect of fructose and glucose on insulin status could not be attributed to differences in body weight. The authors concluded that the fructose moiety appeared to be responsible for the impaired insulin binding and insulin sensitivity produced by sucrose. It has been shown that the infusion of fructose in humans produces a significant decrease in the ATP content of the liver.[99] It was therefore suggested that the impaired insulin binding produced by the feeding of high levels of fructose may be the result of a similar decrease in ATP levels which would precipitate a decrease in cAMP, a metabolite that has been shown to increase the concentration of insulin receptors in cultured human lymphocytes.[100] Whether fructose fed at levels more closely approximating that presently consumed for longer periods of time would produce a similar decrease in insulin binding and sensitivity has yet to be determined.

The effects of fructose feeding on fasting and response levels of blood glucose and insulin in six hospitalized, hypertriglyceridemic men have been reported by Turner et al.[101] Three of the subjects were more than 26% above ideal body weight and two were diabetic. The study was divided into four dietary periods each of about 2 weeks' duration. In two of the dietary periods the subjects were fed a liquid formula diet containing 45% of the calories from carbohydrate (dextromaltose), 40% from fat (one half each of butter fat and corn oil), and 15% from protein (dried milk powder). In the other two dietary periods the fat was replaced by carbohydrate. During one of the periods when the subjects were consuming the fat-containing and fat-free diets, 20% of the dextromaltose was replaced by fructose. Fructose therefore contributed either 9% of the total calories (33 to 46 g per day) in the diet containing fat or 17% of the total calories (90 to 154 g per day) in the diet containing no fat. The diets were given five times throughout the day. The dietary periods were randomized and each subject received each of the diets during the study. The subjects lost an average of 2.1 kg of body weight during the study. No significant changes in plasma glucose or insulin were found after substitution of fructose for dextromaltose in either the fat-containing or fat-free diet. The insulin response to a fasting formula tolerance test (the morning formula containing 20% of the daily total calories) was determined after each dietary period and results expressed as the relative insulin response (mean of insulin responses at 30 and 60 min) − (base line insulin)/(base line insulin) × 100. Table 5 summarizes these results. The replacement of dextromaltose by fructose did not significantly affect the insulin responses. Corresponding data for glucose responses were not reported. These results are consistent with those reported earlier in which it was found that fructose, when consumed as part of a meal containing components that can raise blood glucose levels, does not significantly decrease insulin responses.

The effects of the extended feeding of different amounts of fructose on indices of glucose tolerance in men having an abnormally high insulin response following a glycemic stress (n = 12) and their normoinsulinemic controls (n = 12) have been reported.[102] The subjects were fed a diet containing 43% of the calories from carbohydrate, 42% from fat (P/S ratio = 0.4), and 15% from protein in three meals daily. The diets provided a daily average of 550 mg of cholesterol and 5 g of crude fiber. The diets differed only in the 15% of calories provided by either added fructose or wheat starch (0% fructose, 15% wheat starch; 7.5% each of fructose and wheat starch; 15% fructose, 0% wheat starch). Each of the three diets was consumed by the subjects in a crossover design for 5 weeks. Plasma glucose, insulin, and GIP levels after fasting and a sucrose tolerance test (2 g/kg body weight) were determined. The only significant difference in these parameters observed between the subject groups was

Table 5

RELATIVE INSULIN RESPONSES OF HYPERTRIGLYCERIDEMIC MEN AFTER CONSUMING A FORMULA TOLERANCE TEST CONTAINING 20% OF THE TOTAL CALORIC INTAKE AS FRUCTOSE

Formula test diet	Fructose	Relative insulin response[a] (%)
Fat-containing	—	539 ± 214
	+	540 ± 161
Fat-free	—	932 ± 791
	+	1191 ± 734

Note: Values represent the mean ± S.D. from six subjects.

[a] (Mean of insulin values at 30 and 60 min) − (base line insulin)/(base line insulin) × 100.

Adapted from Turner, J. L., Bierman, E. L., Brunzell, J. D., and Chait, A., *Am. J. Clin. Nutr.*, 32, 1043, 1979.

Table 6

MEAN GLUCOSE, INSULIN, AND GIP RESPONSES 30 TO 120 MIN FOLLOWING A SUCROSE LOAD (2 g/kg BODY WEIGHT) AFTER 12 HYPERINSULINEMIC MEN AND THEIR CONTROLS CONSUMED 0, 7.5, OR 15% OF CALORIES AS FRUCTOSE FOR 5 WEEKS

Dietary fructose (% kcal)	Glucose, all subjects (mg %)	Insulin (μu/mℓ)		GIP, all subjects (pg/mℓ)
		Hyperinsulinemics	Controls	
0	116[a]	146[a]	59[a]	1764[a]
7.5	123[a,b]	152[a]	59[a]	1911[b]
15	125[b]	179[b]	67[b]	2010[b]

Note: Means in a column not sharing a common superscript letter are significantly different from each other ($p < 0.05$).

Adapted from Hallfrisch, J., Ellwood, K. C., Michaelis, O. E., IV, Reiser, S., O'Dorisio, T. M., and Prather, E. S., *J. Nutr.*, 113, 1817, 1983.

for the insulin response. Fasting glucose levels were significantly higher during both the 7.5 and 15% than after the 0% fructose diet, but these differences were very small and well within the normal range. There was no significant effect of fructose feeding on either fasting insulin or GIP levels. Table 6 presents the glucose, insulin, and GIP responses after the sucrose load. The average glucose response 30 to 180 min following the sucrose load was significantly higher after the subjects consumed the 15% than after the 0% fructose diet. Corresponding insulin responses were significantly higher in both the hyperinsulinemic and control subjects after they consumed the 15% than after either the 0 or 7.5% fructose diet. In contrast to the results obtained after the feeding of a load dose of fructose to unadapted subjects,[63,89] the GIP responses to a sucrose load were significantly higher after the subjects consumed either the 7.5 or 15% than after the 0% fructose diet. Fasting levels of plasma thyroxine (but not cortisol) growth hormone, or testosterone, were significantly higher after the subjects consumed the diets containing 7.5 and 15% fructose than after the diet with no added fructose.[103]

It is clear from the studies described in this section that there exist apparently contradictory findings pertaining to the effects of fructose feeding on indices of glucose tolerance. A probable source of these contradictory results can be found in the differences in experimental conditions used in these studies. A comparison of some of the experimental conditions employed by Turner et al.[101] and Hallfrisch et al.[102] in which insulin responses of subjects with abnormally high triglyceride or insulin levels were either not affected[101] or adversely affected[102] by fructose feeding can illustrate this point. In one of the studies[101] the feeding of fructose was compared to dextromaltose which contains 56% maltose and 41% dextrins, while in the other study[102] the feeding of fructose was compared to wheat starch. The glycemic index and resulting insulin response would be expected to be higher after the feeding of dextromaltose than after the feeding of wheat starch.[90] In the study of Turner et al.,[101] the diet was provided in five equal meals. In contrast, in the study of Hallfrisch et al.,[102] the diet was presented in three meals with most of the calories consumed at dinner. It has been shown that consumption of total calories in a gorging pattern as compared to a nibbling pattern adversely affects glucose tolerance in humans.[104,105] Differences in the amount and composition (P/S ratio) of fat in the diet have been shown to influence the ability of fructose-containing sugar to increase blood triglycerides.[94] Increases in blood triglycerides have been shown to reduce the number of insulin-receptors,[74] thereby adversely affecting glucose tolerance. Other experimental factors that could affect differentially the ability of fructose to affect glucose tolerance include the amount and nature of dietary fiber, the length of time the fructose diet is fed, and the trace mineral status of the diet. In view of the inconclusive nature of the comparatively few studies dealing with the effects of fructose feeding on indices of glucose tolerance and the undesirable effects of fructose feeding on other metabolic risk factors associated with disease states (e.g., blood triglycerides and uric acid), the intake of diets containing increasing amounts of fructose does not appear to be warranted.

REFERENCES

1. **Jackson, W. P. U., van Mieghem, W., and Keller, P.,** Insulin excess as the initial lesion in diabetes, *Lancet,* 1, 1040, 1972.
2. **Kraft, J. R. and Nosal, R. A.,** Insulin values and diagnosis of diabetes, *Lancet,* 1, 637, 1975.
3. **Pyörälä, K.,** Relationship of glucose tolerance and plasma insulin to the incidence of coronary heart disease: results from two population studies in Finland, *Diabetes Care,* 2, 131, 1979.
4. **Welborn, T. A. and Wearne, K.,** Coronary heart disease incidence and cardiovascular mortality in Busselton with reference to glucose and insulin concentrations, *Diabetes Care,* 2, 154, 1979.
5. **Ducimetiere, P., Eschwege, L., Papoz, J. L., Richard, J. L., Claude, J. R., and Rosselin, G.,** Relationship of plasma insulin levels to the incidence of myocardial infarction and coronary heart disease mortality in a middle-aged population, *Diabetologia,* 19, 205, 1980.
6. **Stout, R. S.,** The relationship of abnormal circulating insulin levels to atherosclerosis, *Atherosclerosis,* 27, 1, 1977.
7. **Fuller, J. H., Shipley, M. J., Rose, G., Jarrett, R. J., and Keen, H.,** Coronary-heart-disease risk and impaired glucose tolerance, the Whitehall study, *Lancet,* 1, 1373, 1980.
8. **Wahlberg, F. and Thomasson, B.,** Glucose tolerance in ischaemic cardiovascular disease, in *Carbohydrate Metabolism and Its Disorders,* Vol. 2, Dickens, F., Randle, P. J., and Whelan, W. J., Eds., Academic Press, Orlando, Fla., 1968, chap. 8.
9. **Olefsky, J. M., Farquhar, J. W., and Reaven, G. M.,** Reappraisal of the role of insulin in hypertriglyceridemia, *Am. J. Med.,* 57, 551, 1974.

10. **Baron, P., Griffaton, G., and Lowy, R.,** Metabolic inductions in the rat after an intraperitoneal injection of fructose and glucose, according to the nature of the dietary carbohydrate. II. Modifications after seven months of diet, *Enzyme*, 12, 481, 1971.

11. **Bruckdorfer, K. R., Khan, I. H., and Yudkin, J.,** Fatty acid synthetase activity in the liver and adipose tissue of rats fed with various carbohydrates, *Biochem. J.*, 129, 439, 1972.

12. **Vrána, A., Fábry, P., Slabochová, Z., and Kazdová, L.,** Effect of dietary fructose on free fatty acid release from adipose tissue and serum free fatty acid concentration in the rat, *Nutr. Metab.*, 17, 74, 1974.

13. **Waterman, R. A., Romsos, D. R., Tsai, A. C., Miller E. R., and Leveille, G. A.,** Effects of dietary carbohydrate source on growth, plasma metabolites and lipogenesis in rats, pigs and chicks, *Proc. Soc. Exp. Biol. Med.*, 150, 220, 1975.

14. **Kang, S. S., Bruckdorfer, K. R., and Yudkin, J.,** Influence of different dietary carbohydrates on liver and plasma constituents in rats adapted to meal feeding, *Nutr. Metab.*, 23, 301, 1979.

15. **Bird, M. I. and Williams, M. A.,** Triacylglycerol secretion in rats: effects of essential fatty acids and influence of dietary sucrose, glucose or fructose, *J. Nutr.*, 112, 2267, 1982.

16. **Zavaroni, I., Chen, Y.-D. I, and Reaven, G. M.,** Studies of the mechanism of fructose-induced hypertriglyceridemia in the rat, *Metabolism*, 31, 1077, 1982.

17. **Cohen, A. M., Teitelbaum, A., and Rosenmann, E.,** Diabetes induced by a high-fructose diet, *Metabolism*, 26, 17, 1977.

18. **Cohen, A. M.,** Genetically determined response to different ingested carbohydrates in the production of diabetes, *Horm. Metab. Res.*, 10, 86, 1978.

19. **Vrána, A., Fábry, P., and Kazdová, L.,** Liver glycogen synthesis and glucose tolerance in rats adapted to diets with a high proportion of fructose or glucose, *Nutr. Metab.*, 22, 262, 1978.

20. **Sleder, J., Chen, Y.-D. I., Cully, M. D., and Reaven, G. M.,** Hyperinsulinemia in fructose-induced hypertriglyceridemia in the rat, *Metabolism*, 29, 303, 1980.

21. **Zavaroni, I., Sander, S., Scott, S., and Reaven, G. M.,** Effect of fructose feeding on insulin secretion and insulin action in the rat, *Metabolism*, 29, 970, 1980.

22. **Blakely, S. R., Hallfrisch, J., Reiser, S., and Prather, E. S,** Long-term effects of moderate fructose feeding on glucose tolerance parameters in rats, *J. Nutr.*, 111, 307, 1981.

23. **Tobey, T. A., Moden, C. E., Zavaroni, I., and Reaven, G. M.,** Mechanism of insulin resistance in fructose-fed rats, *Metabolism*, 31, 608, 1982.

24. **Vrána, A. and Fábry, P.,** Metabolic effects of high sucrose or fructose intake, *World Rev. Nutr. Diet.*, 142, 56, 1983.

25. **Boot-Handford, R. and Heath, H.,** Identification of fructose as the retinopathic agent associated with the ingestion of sucrose-rich diets in the rat, *Metabolism*, 29, 1247, 1980.

26. **Boot-Handford, R. P. and Heath, H.,** The effect of dietary fructose and diabetes on the rat kidney, *Br. J. Exp. Pathol.*, 62, 398, 1981.

27. **Topping, D. L. and Mayes, P. A.,** The concentration of fructose, glucose and lactate in the splanchnic blood vessels of rats absorbing fructose, *Nutr. Metab.*, 13, 331, 1971.

28. **Cohen, A. M., Teitelbaum, A., Briller, S., Yanko, L., Rosenmann, E., and Shafrir, E.,** Experimental models of diabetes, in *Sugars in Nutrition*, Sipple, H. L. and McNutt, K. W., Eds., Academic Press, Orlando, Fla., 1974, chap. 29.

29. **Vrána, A., Slabochová, Z., Kazdová, L., and Fábry, P.,** Insulin sensitivity of adipose tissue and serum insulin concentration in rats fed sucrose or starch diets, *Nutr. Rep. Int.*, 3, 31, 1971.

30. **Romsos, D. R. and Leveille, G. A.,** Effect of meal frequency and diet composition on glucose tolerance in the rat, *J. Nutr.*, 104, 1503, 1974.

31. **Reiser, S. and Hallfrisch, J.,** Insulin sensitivity and adipose tissue weight of rats fed starch or sucrose diets ad libitum or in meals, *J. Nutr.*, 107, 147, 1977.

32. **Laube, H., Wojcikowski, C., Schatz, H., and Pfeiffer, E. F.,** The effect of high maltose and sucrose feeding on glucose tolerance, *Horm. Metab. Res.*, 10, 192, 1978.

33. **Hallfrisch, J., Lazar, F., Jorgensen, C., and Reiser, S.,** Insulin and glucose responses in rats fed sucrose and starch, *Am. J. Clin. Nutr.*, 32, 787, 1979.

34. **Hallfrisch, J., Cohen, L., and Reiser, S.,** Effects of feeding rats sucrose in a high fat diet, *J. Nutr.*, 111, 531, 1981.

35. **Maugh, T. H., II,** Hormone receptors — new clues in the cause of diabetes, *Science*, 193, 220, 1976.

36. **Leiter, E. H., Coleman, D. L., Ingram, D. K., and Reynolds, M. A.,** Influences of dietary carbohydrate on the induction of diabetes in C57BL/KsJ-db/db diabetes mice, *J. Nutr.*, 113, 184, 1983.

37. **Curry, D. L., Curry, K. P., and Gomez, M.,** Fructose potentiation of insulin secretion, *Endocrinology*, 91, 1493, 1972.

38. **Dunnigan, M. G. and Ford, J. A.,** The insulin response to intravenous fructose in relation to blood glucose levels, *J. Clin. Endocrinol. Metab.*, 40, 629, 1975.

39. Lawrence, J. R., Gray, C. E., Grant, I. S., Ford, J. A., McIntosh, W. B., and Dunnigan, M. G., The insulin response to intravenous fructose in maturity-onset diabetes mellitus and in normal subjects, *Diabetes*, 29, 736, 1980.

40. Cryer, A., Riley, S. E., Williams, E. R., and Robinson, D. S., Effects of fructose, sucrose and glucose feeding on plasma insulin concentrations and on adipose-tissue clearing-factor lipase activity in the rat, *Biochem. J.*, 140, 561, 1974.

41. Jain, K., Logothetopoulos, J., and Zucker, P., The effect of D - and L -glyceraldehyde on glucose oxidation, insulin secretion and insulin biosynthesis by pancreatic islets of the rat, *Biochim. Biophys. Acta*, 399, 384, 1975.

42. Ashcroft, S. J. H., Weerasinghe, L. C. C., and Randle, P. J., Interrelationship of islet metabolism, adenosine triphosphate content and insulin release, *Biochem. J.*, 132, 223, 1973.

43. Hellman, B., Idahl, L.-Å., Lernmark, Å., Sehlin, J., and Taljedäl, I.-B., The pancreatic β-cell recognition of insulin secretagogues, *Arch. Biochem. Biophys.*, 162, 448, 1974.

44. Curtis-Prior, P. B., Trethewey, J., Stewart, G. A., and Hanley, T., The contribution of different organs and tissues of the rat to assimilation of glucose, *Diabetologia*, 5, 384, 1969.

45. Hill, R., Baker, N., and Chaikoff, I. L., Altered metabolic patterns induced in the normal rat by feeding an adequate diet containing fructose as sole carbohydrate, *J. Biol. Chem.*, 209, 705, 1954.

46. Vrána, A., Fábry, P., Kazdová, L., and Zvolánková, K., Effect of the type and proportion of dietary carbohydrate on serum glucose levels and liver and muscle glycogen synthesis in the rat, *Nutr. Metab.*, 22, 313, 1978.

47. Jeffcoate, S. L. and Moody, A. J., The role of the liver in the disposal of orally administered ^{14}C-glucose in the normal rat, *Diabetologia*, 5, 293, 1969.

48. Blumenthal, M. D., Abraham, S., and Chaikoff, I. L., Dietary control of liver glucokinase activity in the normal rat, *Arch. Biochem. Biophys.*, 104, 215, 1964.

49. Zakim, D., Pardini, R. S., Herman, R. H., and Sauberlich, H. E., Mechanism for the differential effects of high carbohydrate diets on lipogenesis in rat liver, *Biochim. Biophys. Acta*, 144, 242, 1967.

50. Hornichter, R. D. and Brown, J., Relationship of glucose tolerance to hepatic glucokinase activity, *Diabetes*, 18, 257, 1969.

51. Adelman, R. C., Spolter, P. D., and Weinhouse, S., Dietary and hormal regulation of enzymes of fructose metabolism in rat liver, *J. Biol. Chem.*, 241, 5467, 1966.

52. Chevalier, M. M., Wiley, J. H., and Leveille, G. A., Effect of dietary fructose on fatty acid synthesis in adipose tissue and liver of the rat, *J. Nutr.*, 102, 337, 1972.

53. Tuovinen, C. G. R. and Bender, A. E., Some metabolic effects of prolonged feeding of starch, sucrose, fructose and carbohydrate-free diet in the rat, *Nutr. Metab.*, 19, 161, 1975.

54. Freedland, R. A. and Harper, A. E., Metabolic adaptations in higher animals. I. Dietary effect on liver glucose-6-phosphatase, *J. Biol. Chem.*, 228, 743, 1957.

55. Cohen, A. M., Briller, S., and Shafrir, E., Effect of long term sucrose feeding on the activity of some enzymes regulating glycolysis, lipogenesis and gluconeogenesis in rat liver and adipose tissue, *Biochim. Biophys. Acta*, 279, 129, 1972.

56. Kaufmann, U. and Froesch, E. R., Inhibition of phosphorylase-a by fructose-1-phosphate, α-glycerophosphate and fructose-1,6-diphosphate: explanation for fructose-induced hypoglycaemia in hereditary fructose intolerance and fructose-1,6-diphosphatase deficiency, *Eur. J. Clin. Invest.*, 3, 407, 1973.

57. Van den Berghe, G., Metabolic effects of fructose in the liver, *Curr. Top. Cell. Regul.*, 13, 97, 1978.

58. Sestoft, L., Fructose and the dietary therapy of diabetes mellitus, *Diabetologia*, 17, 1, 1979.

59. Van de Werve, G. and Hers, H.-G., Mechanism of activation of glycogen phosphorylase by fructose in the liver, *Biochem. J.*, 178, 119, 1979.

60. Johnson, P. R. and Miller, T. B., Jr., Adverse effects of fructose in perfused livers of diabetic rats, *Metabolism*, 31, 121, 1982.

61. Regan, J. J., Jr., Doorneweerd, D. D., Gilboe, D. P., and Nuttall, F. Q., Influence of fructose on the glycogen synthase and phosphorylase systems in rat liver, *Metabolism*, 29, 965, 1980.

62. Burch, H. B., Lowry, O. H., Meinhardt, L., Max, P., Jr., and Chyu, K., Effect of fructose, dihydroxyacetone, glycerol, and glucose on metabolites and related componds in liver and kidney, *J. Biol. Chem.*, 245, 2092, 1970.

63. Sestoft, L., Fructose and health, *Nutr. Update*, 1, 39, 1983.

64. Niewoehner, C. B., Gilboe, D. P., Nuttall, G. A., and Nuttall, F. Q., Metabolic effects of oral fructose in the liver of fasted rats, *Am. J. Physiol.*, 247, E505, 1984.

65. Keen, H. and Chlouverakis, C., Metabolic factors in diabetic retinopathy, in *Biochemistry of the Retina*, Graymore, C. N., Eds., Academic Press, Orlando, Fla., 1965, 123.

66. Heath, H., Kang, S. S., and Philippou, D., Glucose, glucose-6-phosphate, lactate and pyruvate content of the retina, blood and liver of streptozotocin-diabetic rats fed sucrose- or starch-rich diets, *Diabetologia*, 11, 57, 1975.

67. **Heath, H. and Hamlett, Y. C.**, The sorbitol pathway: effect of streptozotocin induced diabetes and the feeding of a sucrose-rich diet on glucose, sorbitol and fructose in the retina, blood and liver of the rat, *Diabetologia*, 12, 43, 1976.
68. **Yanko, L., Michaelson, I. C., and Cohen, A. M.**, Retinopathy in rats with disturbed carbohydrate metabolism following a high-sucrose diet, *Am. J. Ophthalmol.*, 73, 870, 1972.
69. **Scott, P. P., Greaves, J. P., and Scott, M. G.**, Nutritional blindness in cats, *Exp. Eye Res.*, 3, 357, 1964.
70. **Boot-Handford, R. P.**, The pathogenic action of fructose on the vascular system in relation to diabetic microangiopathy, Ph.D. thesis, University of London, England, 1980.
71. **Katzen, H. M. and Schimke, R. T.**, Multiple forms of hexokinase in the rat: tissue distribution, age dependence, and properties, *Proc. Natl. Acad. Sci. U.S.A.*, 54, 1218, 1965.
72. **Froesch, E. R.**, Fructose metabolism in adipose tissue, *Acta Med. Scand. Suppl.*, 542, 37, 1972.
73. **Vrána, A. and Fábry, P.**, Dietary carbohydrates and adipose tissue metabolism, *Prog. Biochem. Pharmacol.*, 8, 189, 1973.
74. **Bieger, W. P., Michel, G., Barwich, D., and Wirth, A.**, Diminished insulin receptors on monocytes and erythrocytes in hypertriglyceridemia, *Metabolism*, 33, 982, 1984.
75. **Randle, P. J., Garland, P. B., Hales, C. N., and Newsholme, E. A.**, The glucose fatty-acid cycle, its role in insulin sensitivity and the metabolic disturbances of diabetes mellitus, *Lancet*, 1, 785, 1963.
76. **Macdonald, I., Keyser, A., and Pacy, D.**, Some effects, in man, of varying the load of glucose, sucrose, fructose, or sorbitol on various metabolites in blood, *Am. J. Clin. Nutr.*, 31, 1305, 1978.
77. **Bohannon, N. V., Karam, J. H., and Forsham, P. H.**, Endocrine responses to sugar ingestion in man, *J. Am. Diet. Assoc.*, 76, 555, 1980.
78. **Reiser, S., Scholfield, D., Trout, D., Wilson, A., and Aparicio, P.**, Effect of glucose and fructose on the absorption of leucine in humans, *Nutr. Rep. Int.*, 30, 151, 1984.
79. **Crapo, P. A., Kolterman, O. G., and Olefsky, J. M.**, Effects of oral fructose in normal, diabetic, and impaired glucose tolerance subjects, *Diabetes Care*, 3, 575, 1980.
80. **Crapo, P. A., Scarlett, J. A., Kolterman, O. G., Sanders, L. R., Hofeldt, F. D., and Olefsky, J. M.**, The effects of oral fructose, sucrose, and glucose in subjects with reactive hypoglycemia, *Diabetes Care*, 5, 512, 1982.
81. **Ellwood, K. C., Michaelis, O. E., IV, Hallfrisch, J. G., O'Dorisio, T. M., and Cataland, S.**, Blood insulin, glucose, fructose and gastric inhibitory polypeptide levels in carbohydrate-sensitive and normal men given a sucrose or invert sugar tolerance test, *J. Nutr.*, 113, 1732, 1983.
82. **Aitken, J. M. and Dunnigan, M. G.**, Insulin and corticoid response to intravenous fructose in relation to glucose tolerance, *Br. Med. J.*, 3, 276, 1969.
83. **Aitken, J. M., Newton, D. A. G., Hall, P. E., and Dinwoodie, A. J.**, The corticoid, insulin and growth hormone responses to intravenous fructose in men and women, *Acta Med. Scand. Suppl.*, 542, 165, 1972.
84. **Akgün, S. and Ertel, N. H.**, A comparison of carbohydrate metabolism after sucrose, sorbitol, and fructose meals in normal and diabetic subjects, *Diabetes Care*, 3, 582, 1980.
85. **Akgün, S. and Ertel, N. H.**, Plasma glucose and insulin after fructose and high-fructose corn syrup meals in subjects with non-insulin-dependent diabetes mellitus, *Diabetes Care*, 4, 464, 1981.
86. **Crapo, P. A., Scarlett, J. A., and Kolterman, O. G.**, Comparison of the metabolic responses to fructose and sucrose sweetened foods, *Am. J. Clin. Nutr.*, 36, 256, 1982.
87. **Bantle, J. P., Laine, D. C., Castle, G. W., Thomas, J. W., Hoogwerf, B. J., and Goetz, F. C.**, Postprandial glucose and insulin responses to meals containing different carbohydrates in normal and diabetic subjects, *N. Engl. J. Med.*, 309, 7, 1983.
88. **Akgün, S. and Ertel, N. H.**, The effects of sucrose, fructose, and high-fructose corn syrup meals on plasma glucose and insulin in non-insulin-dependent diabetic subjects, *Diabetes Care*, 8, 279, 1985.
89. **Ganda, O. P., Soeldner, J. S., Gleason, R. E., Cleator, I. G. M., and Reynolds, C.**, Metabolic effects of glucose, mannose, galactose, and fructose in man, *J. Clin. Endocrinol. Metab.*, 49, 616, 1979.
90. **Jenkins, D. J. A., Wolever, T. M. S., Taylor, R. H., Barker, H., Fielden, H., Baldwin, J. M., Bowling, A. C., Newman, H. C., Jenkins, A. L., and Goff, D. V.**, Glycemic index of foods: a physiological basis for carbohydrate exchange, *Am. J. Clin. Nutr.*, 34, 362, 1981.
91. **Huttunen, J. K., Mäkinen, K. K., and Scheinin, A.**, Turku sugar studies. XI. Effects of sucrose, fructose and xylitol diets on glucose, lipid and urate metabolism, *Acta Odont. Scand.*, 33(Suppl. 70), 239, 1975.
92. **Bossetti, B. M., Kocher, L. M., Moranz, J. F., and Falko, J. M.**, The effects of physiologic amounts of simple sugars on lipoprotein, glucose, and insulin levels in normal subjects, *Diabetes Care*, 7, 309, 1984.
93. **Crapo, P. A. and Kolterman, O. G.**, The metabolic effects of 2-week fructose feeding in normal subjects, *Am. J. Clin. Nutr.*, 39, 525, 1984.
94. **Reiser, S.**, Effect of dietary sugars on metabolic risk factors associated with heart disease, *Nutr. Health*, 3, 203, 1985.

95. **Nikkilä, E. A.,** Influence of dietary fructose and sucrose on serum triglycerides in hypertriglyceridemia and diabetes, in *Sugars in Nutrition,* Sipple, H. L. and McNutt, K. W., Eds., Academic Press, Orlando, Fla., 1974, chap. 26.

96. **Palumbo, P. J., Briones, E. R., Nelson, R. A., and Kottke, B. A.,** Sucrose sensitivity of patients with coronary-artery disease, *Am. J. Clin. Nutr.,* 30, 394, 1977.

97. **Beck-Nielsen, H., Pedersen, O., and Schwartz Sørensen, N.,** Effects of diet on the cellular insulin binding and the insulin sensitivity in young healthy subjects, *Diabetologia,* 15, 289, 1978.

98. **Beck-Nielsen, H., Pedersen, O., and Lindskov, H. O.,** Impaired cellular insulin binding and insulin sensitivity induced by high-fructose feeding in normal subjects, *Am. J. Clin. Nutr.,* 33, 273, 1980.

99. **Bode, C., Schumacher, H., Goebell, H., Zelder, O., and Pelzel, H.,** Fructose induced depletion of liver adenine nucleotide in man, *Horm. Metab. Res.,* 3, 299, 1971.

100. **Thomopoulos, P., Kosmakos, F. C., Pastan, I., and Lovelace, E.,** Cyclic AMP increases the concentration of insulin receptors in cultured fibroblasts and lymphocytes, *Biochem. Biophys. Res. Commun.,* 75, 246, 1977.

101. **Turner, J. L., Bierman, E. L., Brunzell, J. D., and Chait, A.,** Effect of dietary fructose on triglyceride transport and glucoregulatory hormones in hypertriglyceridemic men, *Am. J. Clin. Nutr.,* 32, 1043, 1979.

102. **Hallfrisch, J., Ellwood, K. C., Michaelis, O. E., IV, Reiser, S., O'Dorisio, T. M., and Prather, E. S.,** Effects of dietary fructose on plasma glucose and hormone responses in normal and hyperinsulinemic men, *J. Nutr.,* 113, 1817, 1983.

103. **Hallfrisch, J., Reiser, S., Prather, E. S., and Canary, J. J.,** Relationships of glucoregulatory hormones in normal and hyperinsulinemic men consuming fructose, *Nutr. Res.,* 5, 585, 1985.

104. **Gwinup, G., Byron, R. C., Roush, W., Kruger, F., and Hamwi, G. J.,** Effect of nibbling versus gorging on glucose tolerance, *Lancet,* 2, 165, 1963.

105. **Pringle, D. J., Wadhwa, P. S., and Elson, C. E.,** Influence of frequency of eating low energy diets on insulin response in women after weight reduction, *Nutr. Rep. Int.,* 13, 339, 1976.

Chapter 6

EFFECTS ON OTHER HORMONES

I. INTRODUCTION

Although insulin is the most researched glucoregulatory hormone, there are a number of others which are influenced by fructose ingestion. The effects of fructose consumption on insulin secretion have been reviewed in Chapter 5. In this chapter, the effects of fructose consumption on other hormones involved in carbohydrate metabolism will be discussed. Since research involving fructose consumption by humans is sparse, some examples will be given using the fructose-containing disaccharide, sucrose. Hormones which will be discussed include gastric inhibitory polypeptide (GIP), glucagon, growth hormone, cortisol (or in rats, corticosterone), and thyroid hormones.

II. GIP

GIP, which is primarily secreted in the proximal small intestine,[1] inhibits gastric acid secretion, thus its name. Its release is triggered by the presence of various dietary components. The levels of GIP in the blood have been reported to be elevated after the ingestion of glucose,[2,3] galactose,[4,5] sucrose,[6-8] invert sugar,[6] triglycerides,[9] and an amino acid mixture,[10] but not mannose, fructose,[11] or xylitol.[12] GIP increases insulin secretion by the pancreas when blood glucose levels are elevated.[13,14] It is not secreted after i.v. administration of glucose.[15] While a single-load dose of fructose did not result in stimulation of GIP release,[11] adaptation to long-term dietary intake of fructose appeared to increase the GIP response to a sucrose load.[8]

Ganda et al.[11] compared the GIP responses of eight young, healthy men to i.v. infusions of glucose, mannose, and galactose and 50-g oral doses of glucose, mannose, galactose, and fructose. Serum levels of GIP at fasting and 30 min after the ingestion of sugars are shown in Table 1. The peak (30 min) levels of GIP were significantly greater than corresponding fasting levels after both the glucose and galactose doses ($p < 0.01$). However, neither mannose nor fructose loads produced significant elevations in GIP.

Ellwood et al.[7] compared the GIP response of 24 men given oral doses of sucrose or its monosaccharide components (glucose and fructose); 12 of the men had abnormally high insulin responses to a glycemic stress and the other 12 were age-, height-, and weight-matched normal controls. Either 2 g of sucrose or 1 g each of fructose and glucose (invert sugar) per kilogram body weight was given to the fasting subjects and blood samples were collected at $1/_2$, 1, 2, and 3 hr. While both sucrose and invert sugar produced significant elevations in GIP, there were no significant differences between the two in either normal or hyperinsulinemic men. There were no differences in glucose responses to the different carbohydrates or between the groups, but insulin levels were higher in the hyperinsulinemic men after the sucrose than after the invert sugar (disaccharide effect). Therefore, differences in insulin levels could not be explained by differences in GIP secretion.

GIP responses were significantly greater after adaptation to diets containing sucrose[6] than to diets containing starch in ten men and nine women 35 to 55 years of age. The subjects consumed typical American diets containing either 30% sucrose or starch in a crossover design for 6 weeks. At the end of this period, GIP levels $1/_2$ and 1 hr after a sucrose load were significantly greater after the subjects consumed sucrose than after starch (Table 2). This study shows that dietary adaptation can alter the GIP response.

Fructose adaptation in hyperinsulinemic and normal men also enhanced the responses of

Table 1

SERUM GIP LEVELS AT FASTING AND 30 MIN AFTER 50-g ORAL DOSES OF GLUCOSE, GALACTOSE, MANNOSE, AND FRUCTOSE IN NORMAL SUBJECTS

Sugar	GIP level (pg/mℓ)	
	Fasting	30 min
Glucose	161 ± 31	387 ± 67
Galactose	191 ± 31	420 ± 51
Mannose	152 ± 25	168 ± 24
Fructose	221 ± 67	237 ± 50

Note: Each value represents the mean ± SEM from eight subjects.

Adapted from Ganda, O. P., Soeldner, J. S., Gleason, R. E., Cleator, I. G. M., and Reynolds, C., *J. Clin. Endocrinol. Metab.*, 49, 616, 1979.

Table 2

GIP RESPONSES IN TEN MEN AND NINE WOMEN AFTER ADAPTATION TO DIETS CONTAINING 30% OF CALORIES AS SUCROSE OR STARCH

Time (hr)	GIP response (pg/mℓ)	
	Sucrose	Starch
0	469	460
$^1/_2$	2073	1594[a]
1	2605	2240[a]
2	2547	2293
3	1718	1670

Note: Analysis of variance significant for diet and time ($p < 0.01$).

[a] Starch values significantly lower than corresponding sucrose values at $^1/_2$ and 1 hr.

Adapted from Reiser, S., Michaelis, O. E., IV, Cataland, S., and O'Dorisio, T. M., *Am. J. Clin. Nutr.*, 33, 1907, 1980.

GIP to a glycemic stress.[8] After consuming diets containing 7.5 and 15% fructose for 5 weeks, both hyperinsulinemic and normal men had greater GIP responses to a glycemic stress than after consuming a diet with no added fructose (Table 3). Insulin responses were also significantly increased. These results illustrate that the GIP response to a single load of fructose may not accurately reflect the adaptive effects of fructose, when consumed as part of a diet for extended periods of time, on GIP levels.

III. GLUCAGON

Glucagon, an important regulator of hepatic glucose production, has been measured in response to various carbohydrates. In general, a high-carbohydrate diet decreases the need for hepatic glucose production, and one would expect it to result in decreased levels of glucagon. Tiedgen and Seitz[16] found that feeding a high-carbohydrate diet to rats produced a two- to threefold increase in blood insulin levels, but no increase in blood glucagon levels, thus favoring glucose utilization. When the rats were fed a high-protein diet, the increase in blood glucagon levels was 2.2-fold greater than the increase in insulin levels, favoring hepatic glucose production. The effects of various carbohydrates, including fructose, on the levels of blood glucagon in humans have been reported in a number of studies.[8,17-20]

Bohannon et al.[17] examined glucagon responses of five men and four women to 100-g oral doses of sucrose, glucose, and fructose. Two of these subjects were obese and three had a family history of diabetes. Both sucrose and glucose caused suppression of glucagon, which would be expected as a result of increased blood glucose levels. After fructose, however, there was no suppression of glucagon.

In a similar experiment, Crapo et al.[18] compared the effects on glucagon of 50-g oral doses of dextrose and fructose in 12 noninsulin-dependent diabetics. Although controls and glucose-intolerant subjects were also included in the experiment, the only glucagon values

Table 3
GIP RESPONSES IN 24 MEN
AFTER CONSUMING THREE
LEVELS OF FRUCTOSE FOR
5 WEEKS EACH

Time (hr)	GIP response (pg/mℓ) % of Calories as Fructose		
	0	7.5	15.0
0	370	375	323
$^1/_2$	1264	1382	1546
1	1691	1948	2108
2	2164	2303	2450
3	1937	2012	1934

Note: Responses after 7.5 and 15.0% fructose significantly greater by analysis of variance than after 0% fructose ($p < 0.05$).

Adapted from Hallfrisch, J., Ellwood, K. C., Michaelis, O. E., IV, Reiser, S., O'Dorisio, T. M., and Prather, E. S., *J. Nutr.*, 113, 1819, 1983.

reported were for 12 of the 17 diabetics. Glucose significantly suppressed glucagon at 15, 30, 120, 180 min when compared to corresponding times after fructose. Sucrose was also given to the subjects, and all carbohydrates were given in meals with protein and fat, but no glucagon levels were reported after sucrose or meals.

Effects of the long-term feeding of fructose-containing diets on glucagon response have also been reported.[8,19,20] Hallfrisch et al.[8] found no difference in responses to a sucrose load after men consumed three levels of fructose for 5 weeks each. Reiser et al.[19] measured only fasting glucagon levels at weekly intervals while ten men and nine women consumed diets containing 30% of calories as sucrose or starch for 6 weeks. Although there were no significant overall effects of diet, blood glucagon levels were higher after the subjects consumed sucrose as compared to starch for 5 weeks. Those subjects classified as carbohydrate-sensitive on the basis of triglyceride levels above 150 mg/dℓ did have higher fasting glucagon levels than normal controls (216 vs. 154 pg/mℓ, $p < 0.025$). Elevations in fasting glucagon would be expected to result in higher fasting glucose levels as a result of increased hepatic production.

Turner et al.[20] fed formula diets in which fructose replaced 20% of the dextromaltose calories in diets containing 45 and 85% total carbohydrate to six hypertriglyceridemic men for 2 weeks. Fasting glucagon levels were measured, but no significant differences were observed.

Animal studies have also generally failed to demonstrate an effect of fructose (or sucrose) in the diet on blood glucagon levels.[21-23] Gardner et al.[21] found significantly greater levels of fasting pancreatic glucagon in the carbohydrate-sensitive BHE rat than in the Wistar rat, but feeding a 54% sucrose diet did not result in different levels of glucagon than feeding a 54% starch diet for 12 weeks. Gardner et al.,[22] in another study, fed lean and obese Zucker rats diets containing 54% sucrose or starch either *ad libitum* or in meals for 6 weeks. All groups of rats, especially those fed *ad libitum*, had very high levels of glucagon, but sucrose did not have an effect. Pancreatic cells were isolated from dogs, and hormone responses to various sugars were studied.[18] Fructose failed to alter the release of glucagon from pancreatic

Table 4
GROWTH HORMONE
RESPONSES AFTER AN
ORAL DOSE OF 100 g OF
GLUCOSE, FRUCTOSE, OR
SUCROSE

Time (hr)	Growth hormone (ng/mℓ)		
	Glucose	**Fructose**	**Sucrose**
0	3.4	3.7	4.4
1	2.1	2.0	2.3
2	2.3	1.7	1.9
3	6.2	2.4	2.7
4	10.7[a]	1.9	8.4
5	12.5[a]	3.3	5.3

Note: Either eight or nine subjects per group

[a] Significantly greater than fasting values ($p < 0.05$).

Adapted from Bohannon, N. V., Karam, J. H., and Forsham, P. H., *J. Am. Diet. Assoc.*, 76, 555, 1980.

cells. These studies indicate that fructose probably does not have a direct effect on the release of glucagon.

IV. GROWTH HORMONE

Growth hormone was measured in response to oral doses of sucrose, glucose, and fructose.[17] Nine subjects were given at least two of the three carbohydrate tolerance tests and blood samples were collected for 5 hr (Table 4). There was a late stimulation of growth hormone secretion at 4 and 5 hr after glucose and sucrose, but not fructose. This stimulation would be expected to occur as a result of the rapid fall in plasma glucose.

An experiment studying the effects of exercise on glucoregulatory hormones after the consumption of various carbohydrates in nine male volunteers has been reported.[24] Growth hormone was measured before and after ingestion of 75 g of glucose or fructose and a sweet placebo. Subjects exercised vigorously 45 min after the carbohydrate load in a cycle ergometer at 70% of VO_2 max. Carbohydrate source did not affect growth hormone levels, but exercise for 15 min caused significant increases ($p < 0.05$) which continued up to 30 min of exercise and during the recovery period. This stimulation of growth hormone by exercise cannot be due solely to the rapid fall in blood glucose, since there was very little change in the blood glucose levels of the placebo group and a much greater drop in blood glucose of the glucose than the fructose group.

Adaptation to fructose by the hyperinsulinemic and normal men previously mentioned[8] did not affect fasting growth hormone levels,[25] but fasting levels would not be expected to accentuate any effects of the level of dietary fructose on growth hormone levels, since responses to carbohydrates do not occur until 4 to 5 hr after ingestion.[17] Growth hormone secretion is abnormally controlled in diabetics and prediabetics[26] and is reported to induce insulin resistance.[27] Fructose in the diet or after an oral dose does not appear to have any effect on growth hormone secretion.

Table 5
FASTING LEVELS OF PLASMA THYROXINE OF 12 HYPERINSULINEMIC AND 12 NORMAL MEN CONSUMING THREE LEVELS OF FRUCTOSE

| | Thyroxine level (μg/dℓ) | | | |
| | % of Calories as fructose | | | |
Subjects	0	7.5	15	Group mean
Hyperinsulinemic	7.4	8.2	8.0	7.9[a]
Controls	8.6	9.3	9.4	9.1[b]
Diet mean	8.0[a]	8.7[b]	8.7[b]	

Note: Group or diet means not sharing a common superscript are different ($p < 0.05$).

Adapted from Hallfrisch, J., Reiser, S., Prather, E. S., and Canary, J. J., *Nutr. Res.*, 5, 585, 1985.

V. CORTISOL

Both oral glucose[28] and dietary sucrose[29] have been reported to increase plasma cortisol. Fructose ingestion before and during exercise did not result in cortisol levels different from those due to glucose ingestion or a sweet placebo.[24] However, exercise did cause a 10 to 20% increase in cortisol ($p < 0.02$) regardless of carbohydrate or placebo. This increase is probably secondary to hypoglycemia occurring as a result of exercise. Fasting cortisol was not affected by dietary intake of three levels of fructose in hyperinsulinemic and normal men.[25] Any effect fructose might have on cortisol secretion would probably be secondary to its adaptive effects on insulin secretion. Glucocorticoid excess has been reported to reduce insulin sensitivity.[30] Whether the reverse is true has not been fully examined.

VI. THYROID HORMONES

Thyroxine levels have been reported to be normal in both type I and type II diabetics, but triiodothyroxine levels are reduced.[31] This thyroid abnormality is due to a reduced conversion of thyroxine to triiodothyroxine. Hyperinsulinemic men exhibited significantly ($p < 0.05$) lower fasting thyroxine levels irrespective of the fructose content of the diet (Table 5). Thyroxine levels of both hyperinsulinemic and normal men were found to be increased ($p < 0.05$) 5 weeks after they consumed diets containing 7.5 and 15% of calories from fructose as compared to a diet containing no added fructose. The higher levels of thyroxine observed after feeding increasing levels of fructose may be related to the increased norepinephrine levels obtained after the feeding of simple sugars.[32]

VII. SUMMARY

Interrelationships among the glucoregulatory hormones are very complicated and not fully understood. Little research has been reported describing the effects upon these hormones of either oral fructose doses or adaptation to chronic fructose ingestion in man. Although oral doses of fructose alone do not stimulate GIP, both fructose-containing sucrose and invert sugar do. Adaptations to fructose and sucrose also result in increased GIP responses to a glycemic challenge. Neither a fructose dose nor the longer-term feeding of dietary fructose alters glucagon, growth hormone, or cortisol secretion. Fasting levels of thyroxine increased after adaptation to dietary fructose. Effects of fructose on these hormones as well as on the catecholamines need to be studied more extensively.

REFERENCES

1. **Thomas, F. B., Shook, D. F., O'Dorisio, T. M., Cataland, S., Mekhjian, H. S., Caldwell, J. H., and Mazzaferri, E. L.,** Localization of gastric inhibitory polypeptide release by intestinal glucose perfusion in man, *Gastroenterology,* 72, 49, 1977.
2. **Brown, J. C., Dryburgh, J. R., Ross, S. A., and Dupre, J.,** Identification and actions of gastric inhibitory polypeptide, *Recent Prog. Horm. Res.,* 31, 487, 1975.
3. **Karam, J. H., Grasso, S. G., Wegienka, L. C., Grodsky, G. M., and Forsham, P. H.,** Effect of selected hexoses, of epinephrine and of glucagon on insulin secretion in man, *Diabetes,* 15, 571, 1966.
4. **Ross, S. A. and Dupre, J.,** Effects of ingestion of triglyceride or galactose on secretion of gastric inhibitory polypeptide and on responses to intravenous glucose in normal and diabetic subjects, *Diabetes,* 27, 327, 1978.
5. **Cleator, I. G. M. and Gourlay, R. H.,** Release of immunoreactive gastric inhibitory polypeptide (IR-GIP) by oral ingestion of food substances, *Am. J. Surg.,* 130, 129, 1975.
6. **Reiser, S., Michaelis, O. E., IV, Cataland, S., and O'Dorisio, T. M.,** Effect of isocaloric exchange of dietary starch and sucrose in humans on the gastric inhibitory polypeptide response to a sucrose load, *Am. J. Clin. Nutr.,* 33, 1907, 1980.
7. **Ellwood, K. C., Michaelis, O. E., IV, Hallfrisch, J. G., O'Dorisio, T. M., and Cataland, S.,** Blood insulin, glucose, fructose, and gastric inhibitory polypeptide levels in carbohydrate-sensitive and normal men given a sucrose or invert sugar tolerance test, *J. Nutr.,* 113, 1732, 1983.
8. **Hallfrisch, J., Ellwood, K. C., Michaelis, O. E., IV, Reiser, S., O'Dorisio, T. M., and Prather, E. S.,** Effects of dietary fructose on plasma glucose and hormone responses in normal and hyperinsulinemic men, *J. Nutr.,* 113, 1819, 1983.
9. **Falco, J. M., Crockett, S. E., Cataland, S., and Mazzaferri, E. L.,** Gastric inhibitory polypeptide (GIP) stimulated by fat ingestion in man, *J. Clin. Endocrinol. Metab.,* 41, 260, 1975.
10. **Thomas, F. B., Mazzaferri, E. L., Crockett, S. E., Mekhjian, H. S., Gruemer, H. D., and Cataland, S.,** Stimulation of secretion of gastric inhibitory polypeptide and insulin by intraduodenal amino acid perfusion, *Gastroenterology,* 70, 523, 1976.
11. **Ganda, O. P., Soeldner, J. S., Gleason, R. E., Cleator, I. G. M., and Reynolds, C.,** Metabolic effects of glucose, mannose, galactose, and fructose in man, *J. Clin. Endocrinol. Metab.,* 49, 616, 1979.
12. **Salminen, E., Salminen, S., Porkka, L., and Koivistoinen, P.,** The effects of xylitol on gastric emptying and secretion of gastric inhibitory polypeptide in the rat, *J. Nutr.,* 114, 2201, 1984.
13. **Eckel, R. H., Fugimoto, W. Y., and Brunzell, J. D.,** Gastric inhibitory polypeptide enhanced lipoprotein lipase activity in cultured preadipocytes, *Diabetes,* 28, 1141, 1979.
14. **Walsh, J. H.,** Endocrine cells of digestive system, in *Physiology of the Gastrointestinal Tract,* Vol. 1., Johnson, L. R., Ed., Raven Press, New York, 1981, 59.
15. **Cataland, S., Crockett, S. E., Brown, J. C., and Mazzaferri, E. L.,** Gastric inhibitory polypeptide (GIP) stimulation by oral glucose in man, *J. Clin. Endocrinol. Metab.,* 39, 223, 1974.
16. **Tiedgen, M. and Seitz, H. J.,** Dietary control of circadian variations in serum insulin, glucagon, and hepatic cyclic-AMP, *J. Nutr.,* 110, 876, 1980.
17. **Bohannon, N. V., Karam, J. H., and Forsham, P. H.,** Endocrine responses to sugar ingestion in man, *J. Am. Diet. Assoc.,* 76, 555, 1980.
18. **Crapo, P. A., Kolterman, O. G., and Olefsky, J. M.,** Effects of oral fructose in normal, diabetic, and impaired glucose tolerance subjects, *Diabetes Care,* 3, 575, 1980.
19. **Reiser, S., Handler, H. B., Gardner, L. B., Hallfrisch, J. G., Michaelis, O. E., IV, and Prather, E. S.,** Isocaloric exchange of dietary starch and sucrose in humans. II. Effect on fasting blood insulin, glucose, and glucagon and on insulin and glucose response to a sucrose-load, *Am. J. Clin. Nutr.,* 32, 2206, 1979.
20. **Turner, J. L., Bierman, E. L., Brunzell, J. D., and Chait, A.,** Effect of dietary fructose on triglyceride transport and glucoregulatory hormones in hypertriglyceridemic men, *Am. J. Clin. Nutr.,* 32, 1043, 1979.
21. **Gardner, L. B., Spannhake, E. B., Keeney, M., and Reiser, S.,** Effect of dietary carbohydrate on serum insulin and glucagon in two strains of rats, *Nutr. Rep. Int.,* 15, 361, 1977.

22. **Gardner, L. B., Michaelis, O. E., IV, and Cataland, S.**, Serum insulin and glucagon and hepatic and adipose cyclic-AMP responses of Zucker rats fed carbohydrate diets ad libitum or in meals, *Nutr. Rep. Int.*, 20, 845, 1979.

23. **Hermansen, K.**, Pancreatic D -cell recognition of D -glucose. Studies with D -glucose, D -glyceraldehyde, dihydroxyacetone, D -mannoheptulose, D -fructose, D -galactose, and D -ribose, *Diabetes*, 30, 203, 1981.

24. **Koivisto, V. A., Karonen, S.-L., and Nikkilä, E. A.**, Carbohydrate ingestion before exercise: comparison of glucose, fructose, and sweet placebo, *J. Appl. Physiol.: Respir. Environ. Exercise Physiol.*, 51, 783, 1981.

25. **Hallfrisch, J., Reiser, S., Prather, E. S., and Canary, J. J.**, Relationships of glucoregulatory hormones in normal and hyperinsulinemic men consuming fructose, *Nutr. Res.*, 5, 585, 1985.

26. **Grecu, E. O., Walter, R. M., Jr., and Gold, E. M.**, Paradoxical release of growth hormone during oral glucose tolerance test in patients with abnormal glucose tolerance, *Metabolism*, 32, 134, 1983.

27. **Rizza, K. A., Cryer, P. E., and Gerich, J. G.**, Role of glucagon, catecholamines, and growth hormone in human glucose counterregulation, *J. Clin. Invest.*, 64, 62, 1979.

28. **Stančákcová, A. and Vajó, J.**, The effect of glucose on the metabolism and excretion of cortisol in man, *Horm. Metab. Res.*, 17, 93, 1985.

29. **Yudkin, J. and Szanto, S.**, Increased levels of plasma insulin and eleven hydroxycorticosteroid induced by sucrose and their reduction by phenformin, *Horm. Metab. Res.*, 4, 417, 1972.

30. **Kahn, C. R., Goldfine, I. D., Nevelle, D. M., Jr., and de Meyts, P.**, Alterations in insulin binding induced by changes in vivo in the levels of glucocorticoids and growth hormone, *Endocrinology*, 103, 1054, 1980.

31. **Schlienger, J. L., Ancian, A., Chabrier, G., North, M. L., and Stephan, F.**, Effect of diabetic control on the level of circulating thyroid hormones, *Diabetologia*, 22, 486, 1982.

32. **Welle, S., Lilavivat, U., and Campbell, R. G.**, Thermic effects of feeding in man: increased plasma norepinephrine levels following glucose but not protein or fat consumption, *Metabolism*, 30, 953, 1980.

Chapter 7

LIPOGENESIS AND BLOOD LIPIDS

I. INTRODUCTION

In humans the liver appears to be a primary site of the synthesis of lipid from carbohydrate. A similar type of metabolism appears to exist in the rat, making this species a good animal model with which to study the mechanisms of carbohydrate-induced lipogenesis is relevant to the human. Hepatic triglyceride synthesis and secretion are believed to be the major sources of endogenous triglyceridemia in humans.[1,2] An elevated level of fasting blood triglycerides is generally considered to be a risk factor in the etiology of heart disease.[3-5] Insulin appears to play an important role in hepatic lipogenesis. Numerous studies have shown a relationship between insulin levels, both fasting and in response to a glycemic stress, and blood triglyceride levels.[6-13] It has been proposed that the primary metabolic defect leading to endogenous hypertriglyceridemia in humans is insulin resistance, followed by compensatory hyperinsulinism.[1] The finding of an increased incidence of coronary heart disease in subjects with impaired glucose tolerance[14] suggests a relationship between these diseases that may be associated with increased levels of insulin and triglycerides. The importance of blood triglycerides as a risk factor of heart disease in diabetics has recently been confirmed in a WHO study reported by West et al.[5] which showed that hypertriglyceridemia was the only metabolic variable found to correlate with an increased incidence of ischemic heart disease.

There is almost unaminous agreement among scientists that a correlation exists between cholesterol levels and the development of heart disease.[15] Elevation in the blood cholesterol associated with low-density lipoproteins (LDL) and very low-density lipoproteins (VLDL) is considered to be a risk factor for heart disease.[16] Elevation of cholesterol associated with high-density lipoproteins (HDL) is considered to be a protective factor against heart disease.[17]

In this chapter, the effects of dietary fructose on hepatic lipogenesis, blood triglycerides, and blood cholesterol in both experimental animals and humans will be described and discussed, with special emphasis on interaction with other environmental factors and genetic predisposition.

II. HEPATIC LIPOGENESIS

A. Experimental Animals

Studies using the rat have shown that fructose is more readily converted by the liver into lipogenic substrates than is glucose. This appears to be true even in animals not adapted to the feeding of fructose or fructose-containing sugars. Table 1 presents the results from five studies in which the comparative rate of conversion of equal amounts of radioactive fructose and glucose into lipogenic substrates by the liver of rats previously fed glucose-based diets was determined.[18-22] These studies show that fructose is converted into lipogenic substrates such as fatty acids, glyceride-glycerol, and triglycerides at a rate 1.4 to 18.9 times greater than that of glucose. These findings can be explained by the differences in the basic hepatic metabolism of these sugars, which favors a lipogenic pathway for fructose. Fructose is converted to fructose-1-phosphate by fructokinase, which has an activity 11 times the combined activities of hexokinase and glucokinase in chow-fed rats.[19] Fructose-1-phosphate is then cleaved to dihydroxyacetone phosphate and glyceraldehyde by fructose-1-phosphate aldolase. Further metabolism of glyceraldehyde requires phosphorylation to glyceraldehyde-3-phosphate, oxidative conversion to glyceric acid followed by phosphorylation, or reduction

Table 1
COMPARATIVE RATE OF CONVERSION OF EQUAL AMOUNTS OF
RADIOACTIVE FRUCTOSE AND GLUCOSE INTO LIPOGENIC SUBSTRATES
BY THE LIVER OF RATS PREVIOUSLY FED GLUCOSE-BASED DIETS

Previous diet	Experimental design	n	Substrate measured	Fructose/glucose[a]	Ref.
58% glucose for 2—14 days	Incubation of liver slices with 110 μM of each sugar	5	Fatty acids	1.4	18
Purina® laboratory chow for at least 1 week	Incubation of liver slices with 5 mM of each sugar	8	Fatty acids	4.5	19
Purina® laboratory chow	0.25—0.40 g/100 g body weight of each of the sugars by stomach tube	4	Liver lipids	5.3	20
	Incubation of 100 mg of minced liver with 15.6 mM of each sugar	8	Fatty acids Glyceride-glycerol Triglycerides	3.1 18.9 11.0	21
	Incubation of isolated parenchymal cells with 5 mM of each sugar	5	Triacylglycerol	2.5	22

[a]   Ratio of substrate measured in the presence of equal amounts of either fructose or glucose.

to glycerol followed by phosphorylation by glycerol kinase to α-glycerol phosphate, a direct precursor of triglyceride. This metabolic sequence effectively produces substrates required for lipogenesis and circumvents two important control mechanisms in the conversion of glucose to glycolytic substrates, the reactions catalyzed by hexokinase and phosphofructokinase (the committed step in glycolysis). It would be expected that the fructose moiety of sucrose would produce a similar increase in hepatic lipogenic substrates as compared to glucose-based carbohydrates. It has also been shown that the perfusion of 40 mg/mℓ of fructose, but not glucose, into the livers of rats fed a standard pellet diet increased the incorporation of radioactive oleic acid into VLDL lipids and the secretion of VLDL triglycerides into the perfusate.[23]

A number of studies have reported on the effects of the adaptation of rats to diets containing fructose as compared to glucose on the incorporation of these sugars into metabolites indicative of hepatic lipogenesis. The results of these studies generally support the perferential use of fructose rather than glucose as a source of carbon for the synthesis of lipogenic substrates. In a long-term feeding study in which rats were fed a diet containing either 72% fructose or glucose for 7 months, the profile of liver substrates was determined after the i.p. injection of 2.5 g of both fructose and glucose per kilogram body weight.[24] The fructose-fed animals were determined to be the best adapted to the rapid mobilization of the sugars on the basis of significantly increased levels of α-glycerol phosphate and malate. In another in vivo study, rats adapted for 11 weeks to a diet containing either 72% fructose or glucose were given by stomach tube radioactive and nonradioactive (0.5 g) loads of the sugar to which they had been adapted.[25] The amount of radioactivity incorporated into both the fatty acid and glycerol moieties of liver triglycerides was significantly greater after the fructose than after the glucose load. The results from four studies in which the rate of incorporation of radioactive glucose or fructose into hepatic fatty acids was compared in rats adapted to the feeding of either glucose or fructose are shown in Table 2. In general, the incorporation of radioactive fructose into fatty acids was greater than that of radioactive glucose in rats fed either glucose or fructose. However, this difference was much greater in rats adapted to fructose. In the study of Chevalier et al.,[26] the tenfold greater concentration of glucose (100 mM) as compared to fructose (10 mM) and the inclusion of insulin in the incubation

Table 2
INCORPORATION OF RADIOACTIVE GLUCOSE OR FRUCTOSE INTO FATTY ACIDS BY THE LIVER OF RATS ADAPTED TO THE FEEDING OF EITHER GLUCOSE OR FRUCTOSE

| | | Radioactivity incorporated into fatty acids | | | | |
| | | Glucose-fed | | Fructose-fed | | |
Experimental design	n	[14]C glucose	[14]C fructose	[14]C glucose	[14]C fructose	Ref.
Rats fed diets containing either 58% glucose or fructose for 2—14 days; liver slices incubated with 110 μM of each sugar	5—7	1.1[a]	1.5[a]	0.3[a]	2.0[a]	18
Rats fed diets containing either 70% glucose or fructose for 2 days following a 2-day fast; liver slices incubated with 5 mM of each sugar	8	21.5[b]	46.0[b]	11.1[b]	74.2[b]	19
Rats fed diets containing either 70% glucose or fructose for 3 weeks; liver slices incubated with 100 mM glucose or 10 mM fructose in a medium containing 0.1 μU insulin/mℓ.	5	207[c]	70[c]	231[c]	164[c]	26
Rats fed diets containing either 66% glucose or fructose for 4 weeks; liver slices incubated with 10 mM of each sugar in a medium containing 0.1 μU insulin/mℓ	4	47[c]	109[c]	26[c]	92[c]	27

[a]   % of added [14]C.
[b]   μmol/100 mg wet weight/3 hr.
[c]   nmol/100 mg wet weight/2 hr.

media probably explain the greater incorporation of glucose than fructose into fatty acids. This explanation is supported by the finding that when equal concentrations (10 mM) of glucose or fructose were used under equivalent experimental conditions, fructose incorporation into fatty acids was significantly greater than that of glucose.[27] The in vitro incorporation of fructose into fatty acids appears to be inhibited at concentrations greater than 75 mM, presumably due to the resultant decreases in the level of ATP.[27] The increase in fructose incorporation into fatty acids in rats adapted to fructose feeding may be due to the increased levels of lipogenic enzymes induced by the feeding of fructose as compared to glucose-based carbohydrates. The results from Table 2 also show that the feeding of fructose as compared to glucose reduces the incorporation of glucose into liver fatty acids. In this regard the conversion of glucose into hepatic lipids and its oxidation to CO_2 by liver slices were also found to be significantly reduced after rats were fed a diet containing 20% fructose and 40% as compared to 60% starch for 26 weeks.[28] The reduced oxidation and lipogenesis of glucose by the liver of fructose-fed rats may be due to the decreased activity of glucokinase found in rats fed fructose[19,29] or fructose-containing sugars.[30] The decreased activity of glucokinase may be aggravated by the concurrent increase in the activity of fructokinase observed after the feeding of fructose[26,31] and the resultant decrease in hepatic ATP required for the hepatic phosphorylation of glucose entering the liver. It appears that the relative importance of the liver is increased and that of adipose tissue decreased when rats are adapted to diets containing fructose as compared to glucose.[25-27]

Studies in which various metabolic intermediates of lipogenesis were incorporated into fatty acids in experimental animals adapted to the feeding of fructose as compared to glucose have not produced clear-cut results. The incorporation of intravenously injected radioactive palmitate into liver triglycerides was not significantly different after rats were fed a commercial pellet ration supplemented with either 10% glucose or fructose in the drinking water for 2 to 4 weeks.[32] Rats fed a diet containing either 72% glucose or fructose for 65 days showed no significant differences in the incorporation of intraperitoneally administered radioactive acetate into hepatic fatty acids and triglycerides.[33] The incorporation of radioactive acetate, glycerol, and oleic acid into the lipids of isolated liver slices was not significantly different after rats were fed a commercial laboratory ration supplemented with either 10% glucose or fructose in the drinking water for 6 days.[20] Despite the apparent lack of an effect of fructose adaptation on the incorporation of these metabolites into hepatic lipids, a significant increase in the radioactivity and level of blood triglycerides was found in the rats fed fructose as compared to glucose.[20,32,33] These results indicate that adaptation to fructose stimulates the mobilization and release of triglycerides from the liver and/or inhibits their removal from the blood. These studies also indicate that fructose-induced increases in hepatic lipogenesis are at least partially dependent on lipogenic substrates derived directly from intact fructose. In contrast to the above studies, the i.v. incorporation of radioactive alanine into saponifiable liver lipid was significantly greater 8 days after the meal feeding to rats of diets containing 70% fructose as compared to glucose.[34] The incorporation of radioactive glycerol phosphate into the lipids of liver following both in vitro incubation of homogenates and in vivo i.p. injection was significantly higher after rats were fed diets containing 75% fructose as compared to glucose for 1 to 11 days.[35] Injection of tritiated water and ^{14}C acetate into the hearts of rats fed for 3 weeks a diet containing 66% fructose as compared to glucose resulted in a significantly greater incorporation of both isotopes in liver fatty acids.[27] The incorporation of intraperitoneally administered tritiated water into hepatic fatty acids was significantly greater after mice had consumed either a fat-containing (5% corn oil) or fat-free diet providing 65 or 70% by weight as fructose in comparison to glucose for 14 days.[36] The apparently contradictory results described in this section may be due to a number of experimental variables including the amount of sugar fed; the length of time fed and the mode of feeding (included in the diet or drinking water, meal feeding, or *ad libitum* feeding); whether the animal was fasted or not fasted; the effect of other dietary components; and the type of metabolic intermediate used and how it was administered.

The activities of various hepatic enzymes involved in the conversion of carbohydrate to lipid have been reported to be increased after experimental animals were fed diets containing fructose as compared to equivalent amounts of glucose polymers or glucose. The increased activities of these enzymes appear to be associated with an increased deposition of liver lipid. Table 3 summarizes the results from numerous studies which permit these general conclusions.[19,20,25-28,32,33,35-55] In most of the studies described, rats were the experimental animal of choice. Diets containing high levels of fructose were usually employed; however, increases in lipogenic enzymes were observed at fructose levels as low as 15% of the diet.[52] The response of the lipogenic parameters to fructose feeding was very rapid, in some cases being evident after only 2 days of feeding.[19,45] There appear to be species differences in the lipogenic response to dietary fructose. The mouse[36] appears to react similarly to the rat, while the chicken[46,56] appears to be unaffected. The inability to increase blood triglycerides in the guinea pig through fructose feeding[20] may be attributed to the extensive metabolism

Table 3
EFFECT OF FEEDING FRUCTOSE AS COMPARED TO GLUCOSE OR GLUCOSE-BASED CARBOHYDRATES ON THE ACTIVITIES OF HEPATIC LIPOGENIC ENZYMES AND AMOUNT OF LIVER LIPID IN EXPERIMENTAL ANIMALS

Experimental conditions	Parameter measured	Fructose	Glucose-based	Ref.
Rats fasted 2 days and then refed a diet containing either 70% glucose or fructose ($n = 8$) for 2 days	Triglycerides (μmol/g)	12.7	9.9	19
	Cholesterol (mg/g)	1.9	1.9	
	Phospholipid phosphorus (mg/g)	0.9	1.0	
Rats fed commercial laboratory chow supplemented with either 10% glucose or fructose in the drinking water for 19 days ($n = 5$)	Triglycerides (mg/g)	6.0[a]	5.4	20
Rats fed a diet containing either 72% glucose or fructose for 1—11 weeks ($n = 11$)	Triglycerides (mg)	625	137	25
Rats fed a diet containing either 70% glucose or fructose for 1—2 weeks ($n = 3$—5)	Malic enzyme	48.0[a]	14.4	26
	Citrate lyase (Enzyme activities expressed as nmol substrate metabolized/min/mg protein)	38.0[a]	17.3	
Rats fed a diet containing either 66% glucose or fructose for 3 weeks ($n = 10$)	Lipid (% wet weight)	4.8[a]	3.3	27
Rats fed a diet containing either 60% starch or fructose for 12 weeks ($n = 6$)	Fat (% dry weight)	27.5[a]	22.9	28
Rats fed commercial pellet chow supplemented with either 10% glucose or fructose in the drinking water for 2—4 weeks ($n = 5$)	Triglycerides (mg/g)	10.0	11.9	32
Rats fed a diet containing 72% by weight of either glucose or fructose for 65 days ($n = 15$—18)	Triglycerides (mg/g/100 g body weight)	37.4	44.4	33
Rats fed a diet containing either 75% glucose or fructose for 4 days ($n = 7$)	Triglycerides (mg/g)	25.3[a]	14.0	35
Mice fed a diet containing either 65% glucose or fructose for 14 days ($n = 6$)	Glucose-6-phosphate dehydrogenase	49.8[a]	29.7	36
	Malic enzyme	314[a]	118	
	Acetyl-CoA carboxylase	21.5[a]	15.3	
	Citrate lyase	71.9[a]	27.5	
	Fatty acid synthetase (Enzyme activities are expressed as nmol/mg protein/min)	48.1[a]	24.8	
Rats fed diets containing either 80% glucose or fructose for 26 weeks ($n = 20$)	Fat (g %)	5.9[a]	3.0	37
Rats fed a diet containing either 68% glucose or fructose for 30 days ($n = 8$)	Lipid (mg/g)	42	39	38
	Triglycerides (mg/g wet weight)	16.0	10.7	
	Cholesterol (mg/g)	3.0	3.0	
	Fatty acid synthetase (μmol NADPH utilized/min/g)	5.6[a]	1.8	

Table 3 (continued)
EFFECT OF FEEDING FRUCTOSE AS COMPARED TO GLUCOSE OR GLUCOSE-BASED CARBOHYDRATES ON THE ACTIVITIES OF HEPATIC LIPOGENIC ENZYMES AND AMOUNT OF LIVER LIPID IN EXPERIMENTAL ANIMALS

Experimental conditions	Parameter measured	Fructose	Glucose-based	Ref.
Rats fed a diet containing either 68% cornstarch or invert sugar for 15 weeks (*n* = 18)	Fat (g %)	9.7[a]	5.9	39
Rats fed a diet containing either 50% glucose or fructose for 4 days (*n* = 6—16)	Lipid (mg/100 g body weight)	255[a]	179	40
	Glucose-6-phosphate dehydrogenase	77.5[a]	45.1	
	Malic enzyme (Enzyme activities expressed as μmol NADP converted/min/100 g body weight)	37.8[a]	24.1	
Rats fed a diet containing either 68% glucose or fructose for 50 days (*n* = 8)	Triglycerides (mg)	245[a]	91	41
	Pyruvate kinase (IU/g)	370[a]	143	
	Glucose-6-phosphate dehydrogenase (IU/g)	61.6[a]	29.6	
Rats fed a diet containing either 65% glucose or invert sugar for 10 days (*n* = 6)	Glucose-6-phosphate dehydrogenase	4.2[a]	2.2	42
	6-Phosphogluconate dehydrogenase	1.4[a]	1.0	
	Malic enzyme	3.5[a]	1.8	
	Pyruvate kinase (Enzyme activities expressed as U/g)	100[a]	81	
Rats fed a diet containing either 70% glucose or fructose for 14 days (*n* = 6—7)	Triglycerides (mg/g)	34.3[a]	20.6	43
Baboons fed a diet containing 40% glucose or fructose with no added cholesterol for 1 year (*n* = 6)	Triglycerides (mg/g)	3.7	8.7	44
	Total cholesterol (mg/g)	4.8	4.8	
	Phospholipids (mg/g)	23.1	22.0	
Rats starved for 2 days and then refed a diet containing either 40% glucose or fructose for 2 days (*n* = 8)	Lipid (mg/100 g body weight)	322[a]	260	45
	Glucose-6-phosphate dehydrogenase	42.6[a]	35.1	
	6-Phosphogluconate dehydrogenase	6.6	5.7	
	Malic enzyme	19.2	16.5	
	Pyruvate kinase	108.0[a]	58.6	
	Citrate lyase	5.0[a]	3.8	
	Acetyl-CoA carboxylase	4.4[a]	3.1	
	Fatty acid synthetase (Enzyme activities expressed as U/100 g body weight)	8.1[a]	5.7	
Rats fed a diet containing either 68% glucose or fructose for 26 days (*n* = 10—12)	Fatty acid synthetase	22.8[a]	9.9	46
	Acetyl-CoA carboxylase	10.9[a]	8.0	
	Fatty acid synthesis	2,555[a]	1,324	
	Cholesterol synthesis	41.6	33.5	

Table 3 (continued)
EFFECT OF FEEDING FRUCTOSE AS COMPARED TO GLUCOSE OR GLUCOSE-BASED CARBOHYDRATES ON THE ACTIVITIES OF HEPATIC LIPOGENIC ENZYMES AND AMOUNT OF LIVER LIPID IN EXPERIMENTAL ANIMALS

Experimental conditions	Parameter measured	Fructose	Glucose-based	Ref.
Chicks fed a diet containing either 60% glucose or fructose for 26 days ($n = 6$—12)	Fatty acid synthetase	54.3	53.1	46
	Acetyl-CoA carboxylase	40.8	34.2	
	Fatty acid synthesis	10,278	9,342	
	Cholesterol synthesis (Enzyme activities expressed as μmol substrate converted to product/mg protein; fatty acid and cholesterol synthesis as nmol of ^{14}C acetate incorporated into fatty acids and into digitonin-precipitable sterols/g tissue/hr, respectively)	121	166	
Rats fed a diet containing either 69% cornstarch or fructose for 4 days ($n = 8$)	Lipids (mg/g)	69.8[a]	44.3	47
	Glucose-6-phosphate dehydrogenase (NADPH μmol/min)	20	16	
	Malic enzyme (NADPH μmol/min)	35[a]	15	
Rats fed a diet containing either 70% glucose or fructose (n not stated)	Pyruvate kinase (U/g)	48[a]	25	48
Baboons fed a diet containing 40% glucose or fructose with 0.1% added cholesterol for 17 months ($n = 6$)	Triglycerides (mg/g)	25.6[a]	13.3	49
	Total cholesterol (mg/g)	4.9[a]	2.9	
Rats fed a diet containing either 70% glucose or fructose for 14 days ($n = 5$)	Lipid (mg/g)	75[a]	55	50
Rats fed a diet containing either 66.5% glucose or fructose for 8—10 weeks ($n = 4$)	Triglycerides (μmol/g)	17.6[a]	11.3	51
	Cholesterol (μmol/g)	2.8	2.6	
	Phospholipids (μmol/g)	31.6	29.5	
Rats fed a diet containing either 54% cornstarch or 39% cornstarch and 15% fructose for 3—15 months ($n = 48$)	Glucose-6-phosphate dehydrogenase (U/g protein)	33.8[a]	28.6	52
	Malic enzyme (U/g protein)	18.0	13.4	
Rats fed a diet containing either 70% glucose or fructose for 3 weeks ($n = 8$)	Pyruvate dehydrogenase (mU/g protein)	840[a]	250	53
	Pyruvate dehydrogenase (mU/100 g body weight)	584[a]	138	
Rats fed a diet containing either 62% starch or fructose for 7 weeks ($n = 10$)	Total lipid (mg/g)	54[a]	44	54
Rats fed a diet containing either 62% glucose or fructose for 11 weeks ($n = 5$)	Glucose-6-phosphate dehydrogenase (U/g protein)	72[a]	35	55
	Malic enzyme (U/g protein)	51[a]	17	

[a] Fructose-fed value significantly greater ($p < 0.05$) than corresponding glucose-based-fed value.

of fructose during its intestinal absorption in this species.[57] However, only 15% of fructose was reported to be converted to glucose during its intestinal absorption in the chicken.[58] In baboons, the feeding of fructose as compared to glucose significantly increased liver triglycerides only when the diets also contained 0.1% cholesterol.[44,49]

The lipogenic enzymes increased by fructose feeding, such as fatty acid synthetase, pyruvate kinase, pyruvate dehydrogenase, citrate lyase, and acetyl-CoA carboxylase (Table 3) encompass the key metabolic steps necessary for the conversion of carbohydrate to lipid by mammalian liver. The fatty acid synthetase complex represents a cluster of seven cytosolic enzymes catalyzing the synthesis of fatty acids from acetyl-CoA. The ultimate source of all the carbon atoms of endogenously formed fatty acids is acetyl-CoA. Pyruvate kinase is a regulatory enzyme in the glycolytic formation of pyruvate, a direct precursor of acetyl-CoA. Pyruvate kinase is inhibited by ATP, thus the increase in pyruvate kinase observed after fructose feeding may be due to a concurrent decrease in hepatic ATP.[27,59] The pyruvate dehydrogenase complex catalyzes the conversion of pyruvate to acetyl-CoA in the mitochondria. As with pyruvate kinase, ATP decreases the activity of the pyruvate dehydrogenase complex, thereby suggesting that the activation of this complex by fructose feeding[53] also may be mediated through a decrease in ATP. Acetyl-CoA itself cannot pass out of the mitochondria into the cytosol for conversion to fatty acids. The acetyl-CoA reaches the cytosol indirectly through incorporation into citrate in the mitochondria and subsequent passage of the citrate into the cytosol where citrate lyase in the presence of ATP and CoA cleaves citrate into acetyl-CoA, oxaloacetate, ADP, and P_i. Before the acetyl-CoA can be utilized by the fatty acid synthetase complex, it must be converted to malonyl-CoA in a reaction catalyzed by the acetyl-CoA carboxylase complex. The synthesis of fatty acids from malonyl-CoA requires large supplies of cytoplasmic NADPH for the reduction steps in this process. A major source of NADPH arises from the NADP-dependent oxidation of glucose-6-phosphate in a reaction catalyzed by glucose-6-phosphate dehydrogenase. Another source of NADPH is oxaloacetate formed in the reaction catalyzed by citrate lyase. The oxaloacetate is reduced to malate, which then undergoes a NADP-dependent reaction catalyzed by malic enzyme to yield pyruvate, CO_2, and NADPH. It is interesting to note that insulin activates many of the key enzymes involved in lipogenesis such as fatty acid synthetase, pyruvate dehydrogenase, citrate lyase, and acetyl-CoA carboxylase. A possible mechanism by which fructose feeding increases these enzymes may be through the increased levels of insulin found in fructose-fed rats.[60-65]

The increased amount of liver lipid found in fructose-fed rats (Table 3) appears to be specific for the triglyceride fraction, since no significant differences were found for hepatic cholesterol[19,38,51] or phospholipids[19,51] between rats fed fructose or glucose-based carbohydrates. The increased level of liver triglycerides appears to be the cause of the increased liver weight consistently found in rats fed fructose as compared to glucose.[27,28,36-41,45,46,50-55] The similarity between the effects of fructose and sucrose on hepatic lipogenic enzyme activity and liver lipids indicates that the fructose moiety of sucrose is primarily responsible for sucrose-induced lipogenesis.[20,26,37-42,45,46,66]

B. Humans

A prerequisite for fructose-induced hepatic lipogenesis in humans would appear to be the ability of dietary fructose to enter the liver in appreciable amounts. In man,[67] as in the rat,[68] fructose is poorly metabolized during its intestinal absorption and appears primarily unchanged in the portal blood. A comparison of the activities of the hepatic enzymes initially metabolizing fructose and glucose in humans[69,70] and rats[19,69] is shown in Table 4. In both human and rat liver the capacity of enzymes phosphorylating fructose as measured by fructokinase activity is much greater than that of glucose as measured by the combined activities of glucokinase and hexokinase. These results are compatible with a more rapid

Table 4
COMPARISON OF ACTIVITIES OF THE HEPATIC ENZYMES INITIALLY METABOLIZING FRUCTOSE AND GLUCOSE IN HUMANS AND RATS

| | Humans | | Rats[a] | | |
| | Fructokinase | Glucokinase + hexokinase | Fructokinase | Glucokinase + hexokinase | Ref. |
Enzyme units					
U/g	1.23	0.31	2.20	1.60	69
	(8)[b]	(5)	(8)	(7)	
μmol	27.4	4.9	31.8	2.8	19, 70
substrate phosphorylated/ min/mg protein	(6)	(6)	(8)	(6)	

[a] Rats previously fed a chow diet.
[b] Numbers in parentheses represent the number of livers measured.

Table 5
COMPARISON OF RATES OF CONVERSION OF RADIOACTIVE FRUCTOSE AND GLUCOSE INTO FATTY ACIDS AND CO_2 BY HUMAN AND RAT LIVER

Liver source	Units	Radioactive fructose		Radioactive glucose		Fructose/glucose	
		Fatty acids	CO_2	Fatty acids	CO_2	Fatty acids	CO_2
Humans (n = 4)	μmol of ^{14}C-sugar recovered as me-tabolite/g/90 min	207	3676	24	697	8.6	5.3
Rats[a] (n = 8)	μmol of ^{14}C-sugar incorporated/100 mg/3 hr	16.8	602.4	3.7	170.8	4.5	3.5

[a] Rats previously fed a chow diet.

Adapted from Zakim, D., Pardini, R. S., Herman, R. H., and Sauberlich, H. E., *Biochim. Biophys. Acta*, 144, 242, 1967; and Zakim, D., Herman, R. H., and Gordon, W. C., Jr., *Biochem. Med.*, 2, 427, 1969.

hepatic metabolism of fructose than of glucose. A comparison of the rates of conversion of radioactive fructose and glucose to fatty acids and CO_2 by human and rat liver also shows metabolic similarities (Table 5).[19,70] In both species the rate of conversion of fructose into fatty acids and CO_2 is much more rapid than that of glucose, comparative rates being 4.5 to 8.6 in humans and 3.5 to 5.3 in rats. These results indicate that the rat is an excellent animal model with which to study the characteristics of fructose-induced hepatic lipogenesis in humans.

From the above results it may also be concluded that in humans, as in rats, fructose produces a much greater hepatic lipogenesis than does glucose. This conclusion is supported by the results of a study by Macdonald.[71] Generally labeled ^{14}C of either fructose or glucose in 25 g of the respective sugar was consumed after an overnight fast by healthy men and by men who had experienced a myocardial infarction. Figure 1 shows the mean specific activity of serum triglycerides 3 hr after the fructose and glucose meals in each subject group. There was a tendency toward a greater incorporation of radioactivity into the tri-glycerides of men who had experienced a myocardial infarction than those who had not after the intake of both sugars. There was a much greater incorporation of fructose as compared to glucose into the triglycerides of both subjects groups.

In a similar study, seven men who had both clinical and electrocardiographic evidence

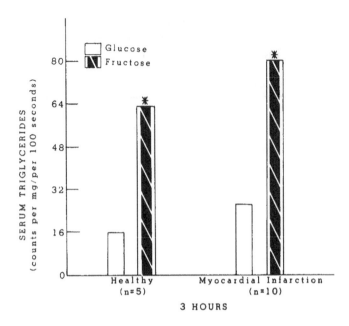

FIGURE 1.   Incorporation of radioactivity into serum triglycerides after healthy men and men who had experienced a myocardial infarction consumed a drink containing radioactive and nonradioactive (25 g) glucose or fructose in a crossover design. (*) Significantly greater than glucose value. (Adapted from Macdonald, I., *Am. J. Clin. Nutr.*, 21, 1366, 1968.)

of a mycoardial infarction were given 50 μCi of either ^{14}C uniformly labeled fructose or glucose together with 25 g of the respective sugar dissolved in 100 mℓ of water in a crossover design separated by about a month.[72] The incorporation of counts from the sugars into the glycerol (Figure 2) and fatty acid (Figure 3) moieties of serum triglycerides was determined over a 4-hr period following the intake of the sugars. The increase in the specific activities of both the glycerol and fatty acid moieties of serum triglycerides was significantly greater 1.5 hr and thereafter following the fructose as compared to the glucose drink. The ratio of fructose to glucose during this time period (1.5 to 4 hr) ranged from 2.8 to 3.7 and 3.5 to 7.0 for the glycerol and fatty acid moieties, respectively.

In a study using i.v. infusions, the incorporation of uniformly labeled ^{14}C fructose into plasma VLDL triglyceride fatty acids of hypertriglyceridemic subjects was consistently higher than that of uniformly labeled ^{14}C glucose after 5 hr.[73] These subjects had been infused with either 10% fructose or glucose for 4 hr prior to and during the study.

III. BLOOD LIPIDS

A. Experimental Animals
1. Blood Triglycerides

Table 6 summarizes the results from 33 studies in which the effect of the feeding of fructose as compared to glucose or starch on blood triglyceride levels in experimental animals was determined.[19,20,25,32,33,35,38,39,41,43,44,46,48-54,62,74-86] In 30 of these studies the experimental animal of choice was the rat. The results generally show that rats fed fructose have significantly higher levels of blood triglycerides than do rats fed glucose or starch. These findings are compatible with the higher hepatic lipogenic capacity of rats fed fructose as compared to glucose-based carbohydrates and with the necessity of transporting larger amounts of neutral fat from the liver to adipose tissue. In this respect it has been shown that the rate of VLDL triglyceride secretion from the liver is 75% greater in fructose- than in glucose-

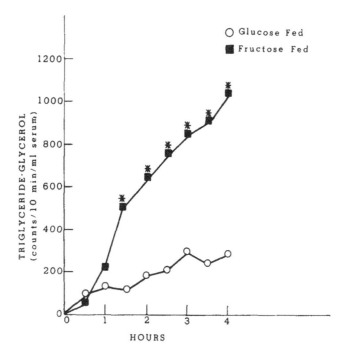

FIGURE 2. Specific activity of the glycerol moiety of serum triglycerides after the ingestion of uniformly labeled and nonradioactive fructose or glucose in men who had experienced a myocardial infarction. Each point represents the mean from seven subjects. (*) Significantly greater than corresponding glucose value ($p < 0.05$). (Adapted from Maruhama, Y., *Metabolism*, 19, 1085, 1970.)

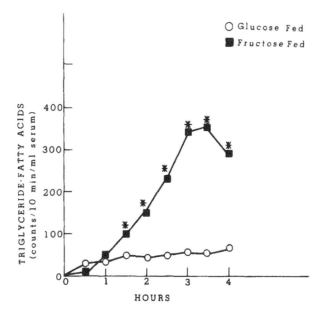

FIGURE 3. Specific activity of the fatty acid moieties of serum triglycerides after the ingestion of uniformly labeled and nonradioactive fructose or glucose in men who had experienced a myocardial infarction. Each point represents the mean from seven subjects. (*) Significantly greater than corresponding glucose value ($p < 0.05$). (Adapted from Maruhama, Y., *Metabolism* 19, 1085, 1970.)

Table 6
**EFFECT OF FEEDING FRUCTOSE AS COMPARED TO GLUCOSE OR
STARCH ON LEVELS OF BLOOD TRIGLYCERIDES IN
EXPERIMENTAL ANIMALS**

Experimental conditions	Fasting blood	Blood triglycerides (mg/100 mℓ)			Ref.
		Fructose	Glucose	Starch	
Rats fasted for 2 days and then refed a diet containing either 70% fructose or glucose for 2 days (*n* = 8)	No	162	104[a]		19
Rats fed commercial laboratory chow supplemented with either 10% fructose or glucose in the drinking water for 19 days (*n* = 5)	Yes	116	46[a]		20
Guinea pigs fed commercial laboratory chow supplemented with either 10% fructose or glucose in the drinking water for 6 days (*n* = 4)	Yes	82	77		20
Rats fed a diet containing either 72% fructose or glucose for 1—11 weeks (*n* = 11)	Yes	86	48[a]		25
Rats fed commercial pellet chow supplemented with either 10% fructose or glucose in the drinking water for 2—4 weeks (*n* = 5)	Yes	142	86		32
Rats fed a diet containing 72% fructose, glucose, or cornstarch for 65 days (*n* 12—18)	Yes	110	18[a]	52	33
Rats fed a diet containing either 75% fructose or glucose for 4 days (*n* = 7)	No	694	643		35
Rats fed a diet containing 68% fructose, glucose, or starch for 30 days (*n* = 8)	No	99	57[a]	54[a]	38
Rats fed a diet containing either 68% invert sugar (34% glucose + 34% fructose) or cornstarch for 15 weeks (*n* = 18)	Yes	937		650[a]	39
Rats fed a diet containing 68% fructose, glucose, or starch for 50 days (*n* = 8)	Yes	78	77	51[a,b]	41
Rats fed a diet containing 70% of calories as either fructose or glucose for 14 days (*n* = 6—7)	No	295	132[a]		43
Baboons fed a diet containing 40% fructose, glucose, or starch with no added cholesterol for 1 year (*n* = 6)	Not stated	129	105	108	44
Rats fed a diet containing either 68% fructose or glucose for 26 days (*n* = 10—12)	No	314	138[a]		46
Chicks fed a diet containing either 60% fructose or glucose for 24 days (*n* = 6—12)	No	91	85		46

Table 6 (continued)
EFFECT OF FEEDING FRUCTOSE AS COMPARED TO GLUCOSE OR STARCH ON LEVELS OF BLOOD TRIGLYCERIDES IN EXPERIMENTAL ANIMALS

Experimental conditions	Fasting blood	Blood triglycerides (mg/100 mℓ)			Ref.
		Fructose	Glucose	Starch	
Rats fed a diet containing either 70% fructose or glucose (time on diet and *n* not stated)	Not stated	350	90		48
Baboons fed a diet containing 40% fructose, glucose, or starch with 0.1% added cholesterol for 17 months (*n* = 6)	Not stated	122	105[a]	96[a]	49
Rats fed a diet containing either 70% fructose or glucose for 14 days (*n* = 5)	No	188	66[a]		50
Rats fed a diet containing either 66.5% fructose or glucose for 8—10 weeks (*n* = 16)	No	96[c]	78[a,c]		51
Rats fed a diet containing either 39% cornstarch and 15% fructose or 54% cornstarch for 3—15 months (*n* = 48)	Yes	152		154	52
Rats fed a diet containing 70% of calories as either fructose or glucose for 3 weeks (*n* = 13)	No	416[c]	168[a,c]		53
Rats fed a diet containing either 62% fructose or cornstarch for 7 weeks (*n* = 10)	Yes	34		37	54
Rats fed a diet containing either 68% fructose or glucose for 1 week (*n* = 11—12)	yes	380	150[a]		62
Young (final body weight = 160 g) rats fed a diet containing 70% fructose, glucose, or starch for 2—3 weeks (*n* = 4—6)	No	112	105		74
Old (final body weight = 507 g) rats fed a diet containing 70% fructose, glucose, or starch for 2—3 weeks (*n* = 4—6)	No	259	173[a]	176[a]	74
Rats fed a diet containing 70% of calories as either fructose or glucose for 3 weeks (*n* = 5—6)	Not stated	204	98[a]		75
Nonobese mice fed a diet containing 64% fructose, glucose, or cornstarch for 24 weeks (*n* = 12)	Yes	86[d]	119	147	76
Obese mice fed a diet containing 64% fructose, glucose, or cornstarch for 24 weeks (*n* = 12)	Yes	108	113	139	76
Rats fed a diet containing 70% of calories as either fructose or glucose for 30 days (*n* = 6)	No	300	170[a]		77
Rats fed a diet containing 70% of calories as either fructose or glucose for 4 weeks (*n* = 15—17)	No	167	87[a]		78
Rats fed a diet containing either 68% fructose or glucose for 36 days (*n* = 6)	Yes	54[c]	65[c]		79

Table 6 (continued)
EFFECT OF FEEDING FRUCTOSE AS COMPARED TO GLUCOSE OR STARCH ON LEVELS OF BLOOD TRIGLYCERIDES IN EXPERIMENTAL ANIMALS

Experimental conditions	Fasting blood	Blood triglycerides (mg/100 mℓ)			Ref.
		Fructose	Glucose	Starch	
Rats fed a diet containing 60% fructose, glucose, or starch for 6 days (n = 7—8)	No	144	80[a]	77[a]	80
Rats fed a diet containing either 70% fructose or glucose for 14 days (n = 5)	No	156	89[a]		81
Rats starved for 4 days and then refed a diet containing either 70% fructose or glucose for 4 days (n = 5)	Yes	117	82[a]		81
Rats fed a fat-free diet containing either 58% fructose or glucose for 3 weeks (n not stated)	No	74	39[a]		82
Rats fed a diet containing 68% fructose, glucose, or starch for 10 weeks (n = 10)	No	172	93	19[a]	83
Rats fed a diet containing either 66% fructose or glucose for 7 days (n = 43 [fructose], 16 [glucose])	Yes	456	242[a]		84
Rats fed a diet containing either 68% fructose or glucose for 8 weeks (n = 10)	No	104[c,e]	57[a,c,e]		85
Normal rats fed a stock diet supplemented with either 4% (w/v) fructose or glucose for 30 days (n = 6—8)	No	115[c]	76[a,c]		86
Streptozotocin-diabetic rats fed a stock diet supplemented with either 4% (w/v) fructose or glucose for 30 days (n = 6—8)	No	151[c]	122[c]		86

[a]   Significantly lower ($p < 0.05$) than corresponding fructose-fed value.
[b]   Significantly lower ($p < 0.05$) than corresponding glucose-fed value.
[c]   Based on a molecular weight of 885 g/mol (triolein).
[d]   Significantly lower ($p < 0.05$) than corresponding starch-fed value.
[e]   Triglycerides associated with lipoprotein density of 1.063 g/mℓ.

fed rats.[82] The precursor relationship between liver and blood triglycerides was illustrated when the increase in blood triglycerides in fructose- as compared to glucose-fed rats was eliminated due to the inhibition of triglyceride release from the liver by orotic acid.[81]

Increases in endogenous rather than exogenous blood triglycerides are associated with the feeding of fructose to rats. In many of the studies outlined in Table 6, the rats were not fasted prior to obtaining blood samples, therefore making it difficult to separate increases in exogenous from endogenous blood triglycerides. However, the lipid content of the diet used in studies where nonfasting blood samples were obtained was usually very small, often 5% or less by weight of the diet.[35,46,50,51,74,80-83,85] Under these conditions the levels of exogenous triglycerides due to the *ad libitum* feeding of the diets would be expected to be very small. In addition, the body weights of the nonfasted rats gave no indication that the fructose-fed rats consumed more diet than the glucose- or starch-fed rats.[38,46,53,74,81,84] In one study the level of fasting triglycerides was found to be higher than that observed in the postprandial state.[39] It may therefore be concluded that, for the most part, the triglyceride levels of nonfasted rats shown in Table 6 represent those endogenously produced.

In many of the studies reported in Table 6, the increases in blood triglycerides in rats observed after fructose feeding were of the same magnitude as those observed after sucrose feeding.[38,39,41,46,51,74,83] These findings are consistent with the contention that sucrose-induced increases in blood triglycerides in the rat are due to the fructose moiety of sucrose.[66]

Several studies have shown that modification of the experimental conditions can influence fructose-induced increases in blood triglycerides in the rat. Rats fed a diet adequate in essential fatty acids and containing 66.5% fructose as compared to glucose showed significant increases in plasma triglycerides.[51] However, when the diets were deficient in essential fatty acids there was no difference in plasma triglycerides of rats fed either fructose or glucose.[51] It appears that fructose levels in excess of 15% of the diet are necessary to elicit a significant increase in blood triglycerides.[52] The feeding of 62% fructose as compared to cornstarch was shown to produce a significant increase in plasma triglycerides only when the rats were copper-deficient.[54] The feeding of diets containing 70% fructose as compared to glucose produced significant increases in serum triglycerides in old rats (final weight 507 g) but not in young rats (final weight 160 g).[74] These results emphasize the importance of evaluating the experimental conditions used to reconcile apparently contradictory findings, a situation which is of special importance in human studies where experimental conditions tend to vary much more than in rat studies.

Table 6 also includes the results of studies in which the experimental animals employed were the guinea pig,[20] baboon,[44,49] chick,[46] and mouse.[76] The failure of the guinea pig to show a significant increase in blood triglycerides when fed fructose as compared to glucose is attributed to the extensive intestinal metabolism of fructose during its absorption in this species.[57] The lack of a fructose-induced increase in plasma triglycerides in chicks is consistent with the failure of fructose to increase hepatic lipogenic enzyme activity in this animal model.[46] The significant increase in hepatic lipogenic enzymes reported in the mouse fed fructose as compared to glucose[36] appears inconsistent with the failure of fructose as compared to glucose to increase plasma triglycerides in this species.[76] Differences in mouse strain, experimental conditions, or an efficient clearance of endogenous triglycerides from the circulation may explain these apparently contradictory findings. Baboons fed a diet containing 40% fructose as compared to glucose or starch, with or without the addition of 0.1% cholesterol, showed increases in serum triglycerides.[44,49] In the study without added cholesterol,[44] the order of ranking of aortas for severity of sudanophilia was fructose, starch, glucose. In the study with the added cholesterol,[49] the ranking for severity of aortic sudanophilia was starch, glucose, fructose. In this latter study, gross atheromatous lesions were seen in three of six baboons fed fructose, one of six fed starch, and none of six fed glucose.

2. Blood Cholesterol

Table 7 presents the results from studies in which blood cholesterol levels were compared in experimental animals fed diets containing fructose as compared to glucose or starch.[19,24,37,39,44,46,49-52,54,76,86,87] In contrast to the effects on fasting blood triglycerides, the action of dietary carbohydrates on blood cholesterol is usually small. This is confirmed by the results in Table 7; in only 2 of the 14 studies summarized was a significant difference in blood cholesterol levels reported after the feeding of the different carbohydrate sources in the context of an apparently nutritionally complete diet. In these two studies using the rat as the experimental animal, an increase in blood cholesterol was observed after the feeding of fructose as compared to glucose[24,50] or starch.[24] The basis for the fructose-induced increase in these studies when compared to studies in which no effect of fructose was observed is not readily apparent. In contrast to copper-supplemented diets, it was reported

Table 7

**EFFECT OF FRUCTOSE AS COMPARED TO GLUCOSE OR
STARCH ON LEVELS OF BLOOD CHOLESTEROL IN
EXPERIMENTAL ANIMALS**

| Experimental conditions | n | Blood cholesterol (mg/100 mℓ) | | | Ref. |
		Fructose	Glucose	Starch	
Rats fasted for 2 days and then refed a diet containing either 70% fructose or glucose for 2 days	8	85	93		19
Rats fed a diet containing 72% fructose, glucose, or tapioca starch for 7 months	14	91	70[a]	67[a]	24
Rats fed a diet containing either 80% glucose or fructose for 26 weeks	20	118	105		37
Rats fed a diet containing either 68% invert sugar or cornstarch for 15 weeks	18	68		64	39
Baboons fed a cholesterol-free diet containing 40% fructose, glucose, or starch for 1 year	6	162	151	156	44
Rats fed a diet containing either 68% fructose or glucose for 26 days	10—12	161	151		46
Chicks fed a diet containing either 60% fructose or glucose for 24 days	6—12	182	196		46
Baboons fed a diet containing 40% fructose, glucose, or starch with 0.1% cholesterol added for 17 months	6	144	155	155	49
Rats fed a diet containing either 70% fructose or glucose for 14 days	5	147	120[a]		50
Rats fed a diet containing either 66.5% fructose or glucose for 8—10 weeks	4	101	139		51
Rats fed a diet containing either 39% cornstarch and 15% fructose or 54% cornstarch for 3—15 months	48	113		119	52
Rats fed a copper-supplemented diet containing either 62% fructose or cornstarch for 7 weeks	10	71		70	54
Rats fed a copper-deficient diet containing either 62% fructose or cornstarch for 7 weeks	9—10	202		113[a]	54
Nonobese mice fed a diet containing 64% fructose, glucose, or cornstarch for 24 weeks	12	194	201	186	76
Obese mice fed a diet containing 64% fructose, glucose, or cornstarch for 24 weeks	12	293	289	265	76
Normal rats fed a stock diet supplemented with either 4% (w/v) fructose or glucose for 30 days	6—8	73	75		86

Table 7 (continued)
EFFECT OF FRUCTOSE AS COMPARED TO GLUCOSE OR STARCH ON LEVELS OF BLOOD CHOLESTEROL IN EXPERIMENTAL ANIMALS

		Blood cholesterol (mg/100 mℓ)			
Experimental conditions	n	Fructose	Glucose	Starch	Ref.
Streptozotocin-diabetic rats fed a stock diet supplemented with either 4% (w/v) fructose or glucose for 30 days	6—8	74	85		86
Rats fed a diet containing either 50% fructose or glucose for 28 days	8	62	63		87

ᵃ Significantly lower ($p < 0.05$) than corresponding fructose value.

that in diets deficient in copper the feeding of fructose produced significantly higher levels of plasma cholesterol than did the feeding of cornstarch.[54,55] However, in both of the studies reporting an increase of blood cholesterol after fructose feeding, the salt mix added to the diets contained copper.[24,50] Since the nature and level of dietary fat mainly determine the level of blood cholesterol, it was of interest to examine the fat content of diets used in those studies in which an effect of fructose was reported in comparison to those in which no effect was found. In one study the diet contained 3% fat as lard.[24] In the other study 0.1% corn oil was used.[50] The amount and nature of the fat are not different from those employed in studies reporting no fructose-induced increase in blood cholesterol. The length of time during which the diets were fed and the age of the rats used also could not explain differences in the effect of fructose. However, differences in the effect of fructose due to the strain of rat used in the studies cannot be eliminated.

In studies with the baboon, not only were total cholesterol levels not affected by the nature of the dietary carbohydrate, but also the percentage of cholesterol associated with LDL was essentially the same (66, 63, and 64% for fructose, glucose, and starch, respectively).[44] Addition of 0.1% cholesterol to the diets produced a similar distribution of cholesterol in the LDL plus VLDL fractions (64, 58, and 58% for fructose, glucose, and starch, respectively).[49] The ratio of high- to low-density serum lipoprotein cholesterol was 0.59 for fructose and 0.72 each for glucose and starch. This comparatively unfavorable distribution of cholesterol found in the lipoproteins of the baboons fed fructose as compared to glucose or starch may partially explain the more numerous aortic atheromatous lesions found in the fructose-fed animals.[49]

Dietary cholesterol at levels of 1 to 3% appears to act synergistically with dietary sucrose, but not dietary starch, to increase total blood cholesterol in experimental animals such as the rat, chicken, and rabbit.[66] The addition of 0.1% cholesterol to the diets of baboons fed 40% fructose or sucrose as compared to starch did not produce an increase in total blood cholesterol levels;[49] therefore, either baboons are insensitive to the interaction between dietary cholesterol and sucrose (and presumably fructose), or cholesterol levels in excess of 0.1% are required to show this interaction in this species.

B. Humans
1. Blood Triglycerides
On the basis of the similarities in hepatic lipogenesis observed between rats and humans and the greater hepatic lipogenic capacity observed in humans when fructose as compared to glucose is administered under a variety of experimental conditions,[70-73] one of the major

Table 8

MEAN VALUES OF FASTING SERUM TRIGLYCERIDES OF MEN AND PRE- AND POSTMENOPAUSAL WOMEN BEFORE AND AFTER CONSUMING FAT-FREE DIETS CONTAINING THE INDICATED CARBOHYDRATE MIXTURES FOR 4 AND 5 DAYS

	Serum triglycerides (mg/100 ml)		
Diet	Men ($n = 5$)	Premenopausal women ($n = 6$)	Postmenopausal women ($n = 3$)
Self-selected (initial)	62 ± 6	84 ± 7	82 ± 8
40% fructose, 60% cornstarch	90 ± 7[a]	88 ± 12	189 ± 16[a]
40% glucose, 60% cornstarch	59 ± 8	69 ± 10	99 ± 11
40% fructose, 60% glucose	94 ± 10[a]	75 ± 10	145 ± 11[a]

Note: Each value represents the mean $\pm$ SEM from the number of subjects shown in parentheses.

[a]   Significantly greater than initial value ($p < 0.01$).

Adapted from Macdonald, I., *Am. J. Clin. Nutr.*, 18, 369, 1966.

concerns relevant to the chronic consumption of fructose in humans is its effect on blood triglyceride levels. This concern is more pertinent to certain population groups which appear to be more sensitive to the lipogenic effects of fructose as compared to glucose-based carbohydrates than to the population as a whole, since these subject groups include those in which fructose intake has been recommended either as a means of weight control or for proposed beneficial effects on glucose tolerance. In this section, a detailed description of the results from studies of the effects of extended fructose feeding on blood triglycerides in humans will be presented. In some of these studies sucrose has been used as the basis of comparison for the effects of fructose on blood triglycerides. Since it appears likely that the fructose moiety of sucrose is primarily responsible for sucrose-induced increases in blood triglycerides in both rats and humans, these studies yield limited metabolic information. However, from the more numerous human studies describing the effects of sucrose as compared to starch on blood triglycerides,[66] it may be possible to predict the environmental conditions producing the more pronounced effects of fructose on blood triglycerides.

The effects of fructose on blood triglycerides appear to depend on the age and sex of the subject. Five men 21 to 28 years of age, six women 21 to 31 (premenopausal), and three women 58 to 68 (postmenopausal) consumed a fat-free diet consisting of 50 g of calcium caseinate, vitamin tablets, and 7.5 g per kilogram body weight of three different carbohydrate mixtures for 5 days.[88] The carbohydrate mixes used were (1) 40% fructose, 60% cornstarch; (2) 40% glucose, 60% cornstarch; and (3) 40% fructose, 60% glucose. Fasting blood triglyceride levels were determined on the day each diet was started and on the 4th and 5th days of the diet. The results are presented in Table 8. In the men and postmenopausal women, serum triglycerides increased significantly over initial levels when fructose was consumed either with cornstarch or glucose. In contrast, glucose produced no significant increases in serum triglycerides in these subjects. Premenopausal women showed no increase in triglycerides after consuming any of the experimental diets. A similar pattern of response of fasting triglycerides as a function of the age and sex of the diet has been reported for sucrose. In contrast to those of men and postmenopausal women,[13,89] blood triglycerides of young women did not increase after they consumed diets containing as much as 70% of the total calories a sucrose.[13,90] In this study,[88] levels of dietary carbohydrate greatly in excess of those presently consumed were fed for a relatively short time period. The use of a fat-free diet also contributed to the distorted dietary pattern. These experimental defects prevent

Table 9
MEAN PLASMA TRIGLYCERIDES OF SUBJECTS WITH ENDOGENOUS HYPERTRIGLYCERIDEMIA AND UNTREATED ADULT-ONSET DIABETES AFTER CONSUMING DIETS CONTAINING STARCH, FRUCTOSE, OR SUCROSE FOR 10—20 DAYS

Subject groups	Triglycerides (mg/100 mℓ)		
	Starch	Fructose	Sucrose
Endogenous hypertriglyceridemic (type IIb or IV) (n = 5)	220	229	270[a]
Untreated adult-onset diabetic (n = 6)	177[b]	201[b]	—

Note: Each value represents the mean of 5—10 measurements from each subject during each period.

[a] Significantly greater ($p < 0.01$) than corresponding starch and fructose values.

[b] Based on a molecular weight of 885 g/mol (triolein).

Adapted from Nikkilä, E. A., in *Sugars in Nutrition,* Sipple, H. L. and McNutt, K. W., Eds., Academic Press, Orlando, Fla., 1974, chap. 26.

a definitive interpretation of these findings as they pertain to present dietary intakes. However, the premise that the intake of high levels of any dietary component will produce adverse metabolic effects is not necessarily true, as seen by the failure of equally high levels of glucose to increase blood triglycerides in the men and older women.

A comparison of the effects of fructose, glucose, sucrose, and starch on serum triglycerides of normal and hypertriglyceridemic subjects was reported by Kaufmann et al.[91] The diets contained an average of 18% protein, 5% fat, and 77% carbohydrate and were usually fed for 2 to 4 weeks. When the sugars were fed, 300 g of sucrose, glucose, or fructose were given per day at the expense of starch. In general, the feeding of fructose as compared to either starch or glucose significantly increased serum triglycerides in subjects with a marked hypertriglyceridemia (above 300 mg %). Fructose appeared to be more effective than sucrose in increasing triglycerides in these subjects. In contrast, subjects with blood triglyceride levels below 200 mg % showed no increase in serum triglycerides when fed fructose as compared to either starch or glucose. The differences in response to fructose exhibited by the hypertriglyceridemic as compared to the normal subjects could not be attributed to differences in the rate of the intestinal absorption of fructose.[92] The use of diets containing such high levels of carbohydrate and low levels of fat makes the relevance of these findings, as they pertain to the usual dietry conditions, questionable also.

A comparison of the effects of feeding diets containing starch, fructose, or sucrose on the plasma triglycerides of subjects with either endogenous hypertriglyceridemia (type IIb or IV) or untreated adult-onset diabetes has been reported.[93] During the period when starch was consumed, the diet contained 45% of calories as carbohydrate (of which 80% was derived from starch), 35% as fat (mainly saturated), and 20% as protein. During the fructose and sucrose periods the subjects were fed the same basic diet, but with 75 to 90 g of the starch replaced by fructose or sucrose. Each dietary period was for 10 to 20 days. The results are shown in Table 9. There was no significant difference between the plasma triglycerides of the subjects with endogenous hypertriglyceridemia during the starch and fructose periods.

However, triglyceride levels were significantly higher ($p < 0.01$) during the sucrose period as compared to either the starch or fructose periods. A turnover study performed at the end of each dietary period indicated that sucrose increased the production rate of plasma triglycerides without affecting the removal efficiency. In the group of subjects with untreated adult-onset diabetes, fructose feeding significantly increased the triglycerides in the two subjects with levels above 200 mg %. In the four subjects with triglyceride levels below 200 mg %, values were essentially the same during the fructose and starch periods. The triglyceride levels during the sucrose period are not shown, since two of the subjects refused to eat the sucrose diet. The dietary conditions employed in this study are reasonably close to those expected in subjects consuming a self-selected diet. Moreover, the time period used was long enough to account for short-term adaptive responses to the diets. It appears, however, that the triglycerides were determined in the nonfasting state. Under these conditions it would be difficult to distinguish between plasma triglycerides derived from exogenous intake of dietary fat and those derived from endogenous synthesis from carbohydrate sources, and possible effects on endogenous triglycerides could be masked by differences in the time and amount of diet consumed by the subjects prior to obtaining the blood sample.

The effects of fructose, sucrose, and xylitol on plasma triglycerides were evaluated in a 2-year study conducted in Finland.[94] Volunteers were divided into three groups and instructed to use fructose ($n = 35$), sucrose ($n = 33$), or xylitol ($n = 48$) as the sole sweetener in their otherwise self-selected diets. The average daily intake of the sweeteners ranged from 56 to 69 g for fructose, 59 to 73 g for sucrose, and 43 to 50 g for xylitol. The lower levels of the sweeteners were observed during the final 8 months of the study. Fasting plasma triglycerides were measured after the 11th, 14th, 18th, and 22nd months of the study. No significant differences were found during any period between the dietary groups. Since initial plasma triglycerides were obtained in the postprandial state, it was not possible to identify subjects with endogenous hypertriglyceridemia and evaluate the effects of diet in this subject group. The experimental design employed in this study had the following deficiencies: (1) no crossover of subjects between diets, (2) comparison of the effects of the sweeteners for subjects consuming self-selected diets of different composition, and (3) poor control of sweetener intake.

The effect on serum triglycerides of feeding a diet containing 2 g per kilogram body weight of fructose as compared to glucose for 4 days to 28 men with diagnosed coronary artery disease has been determined.[95] These men had initial fasting triglycerides in excess of 150 mg %. The diets contained an average of 62% of the calories as carbohydrate, 22% as fat, and 16% as protein, and provided an average of 182 mg of cholesterol daily. In the 14 men fed the diet containing glucose there were no significant differences in serum triglycerides between initial values and those obtained after 4 days on the experimental diet. In contrast, triglyceride levels were significantly higher ($p < 0.01$) after (193 mg %) than before (140 mg %) the other 14 men consumed the diet containing fructose. The clinical significance of the fructose-induced increase in blood triglycerides in these subjects should be evaluated after the feeding of lower levels of fructose for more extended time periods, in diets more closely simulating those presently consumed.

A comparison of the effects on blood triglycerides of feeding diets containing fructose or sucrose for 2 weeks in a crossover design to subjects free of overt disease has been reported.[96] The experimental diets were based on the composition of the self-selected diets of the subjects — 35 to 49% of calories from carbohydrate, 35 to 45% from fat (P/S ratio = 0.5 to 1.5), and 12 to 20% from protein — and provided less than 10 g/day of fiber. The sugars comprised 33% of the carbohydrate calories and ranged from 50 to 107 g/day. No differences in fasting triglycerides were observed after the subjects consumed either sugar for 1 or 2 weeks. From these and other negative results, the investigators concluded that they would not suggest the preferential use of either sugar to apparently healthy individuals. This conclusion may be

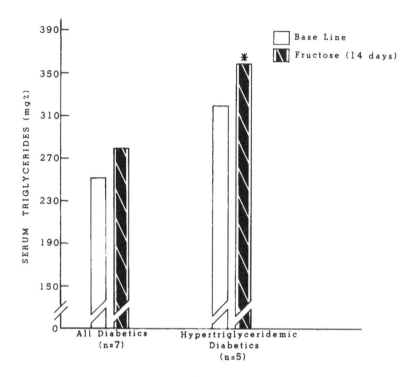

FIGURE 4. Mean fasting serum triglycerides during the base line period and after 14
days of fructose substitution for sucrose in noninsulin-dependent diabetics ($n = 7$) and
those diabetics with initial fasting triglyceride levels in excess of 150 mg % ($n = 5$).
(*) Significantly greater than baseline value ($p < 0.05$). (Adapted from Crapo, P. A.,
Kolterman, O. G., and Henry, R. R., *Diabetes Care*, 9, 111, 1986.)

too general considering the subject group used in the study. The eight subjects ranged in
age from 22 to 32 years and included four women. From previous findings,[13,88,90] it would
be expected that this subject group would be relatively unresponsive to the effects of dietary
fructose or sucrose on blood triglycerides, and therefore would not necessarily reflect the
effect of these sugars on triglycerides in healthy subject groups that are older or comprised
mainly of men.

The levels of blood triglycerides have been determined in both normal subjects and those
with noninsulin-dependent diabetes after they consumed a diet containing sucrose for 3 to
4 days (base line period) followed by a diet in which fructose replaced the sucrose for an
additional 2 weeks.[97,98] During the studies each subject consumed a weight-maintaining diet
that contained 55% of calories from carbohydrate (24% being provided by either sucrose or
fructose), 30% from fat (P/S ratio of about 1), and 15% from protein. The diets provided
300 mg of cholesterol and 25 g of dietary fiber daily. The daily range of sucrose or fructose
provided by the diets was 63 to 115 g. There was no significant difference in fasting serum
triglycerides in the normal subjects after they consumed sucrose (base line) or fructose for
2 weeks.[97] Figure 4 presents the fasting serum triglycerides obtained with the seven diabetic
subjects.[98] There was no significant difference between serum triglyceride levels obtained
after the base line period and 2 weeks after fructose feeding in the subject group as a whole.
However, in five of the seven subjects with preexisting triglyceride levels above 150 mg
%, there was a significant ($p < 0.05$) increase of 13% between values obtained at base line
and those obtained after 2 weeks of fructose feeding. These results again illustrate the
difficulties inherent in drawing general conclusions regarding the effects of dietary components on metabolic risk factors based on overall results from heterogeneous population
groups. Moreover, even within a population group with a common metabolic defect such

as type II diabetes, differences in response to dietary components exist in the individual subjects.

The effect on fasting plasma triglyceride levels of the replacement of glucose-based carbohydrates by fructose in humans exhibiting defects in carbohydrate metabolism has been the subject of two relatively recent human studies.[99,100] Since these studies yielded apparently contradictory results with subjects exhibiting varying degrees of hypertriglyceridemia, it is of interest to compare the experimental conditions employed. In one study,[99] six hospitalized men with type IV hyperlipoproteinemia (pretest fasting triglycerides 207 to 1583 mg %) were fed four different liquid formula diets each for about 2 weeks in a randomized crossover pattern. Two of the diets contained 45% of the calories as carbohydrate (dextromaltose), 40% as fat (one half butter fat, one half corn oil), and 15% as protein (dried milk powder). In the other two diets carbohydrate replaced all the fat (85% of the calories as carbohydrate, 15% as protein). During one of the periods when the subjects were consuming the fat-containing and -free diets, 20% of the dextromaltose was replaced by fructose. Fructose therefore contributed either 9% (33 to 46 g per day) or 17% (90 to 154 g per day) of the calories in the fat-containing and -free diets, respectively. The diets were given five times throughout the day. Mean levels of fasting triglycerides were 238 mg % higher when the subjects consumed the fat-free, high-carbohydrate diets than when they consumed the fat-containing, lower-carbohydrate diets, but this change was not significant. The inclusion of fructose into either the fat-containing or -free diet did not significantly affect the levels of fasting plasma triglycerides. A significant 16 to 21% decrease in triglyceride turnover was found in the fat-free diet when fructose replaced a portion of the dextromaltose. Since plasma triglyceride levels showed no decrease due to fructose in the fat-free diet, the reduction of triglyceride turnover was attributed to a reduction in the peripheral disposal of triglycerides. In the other study,[100] the effect of replacing wheat starch with fructose on fasting plasma triglycerides of 12 men having an abnormally high insulin response following a glycemic stress and of 12 normoinsulinemic men was determined. The subjects were fed a basic diet containing 43% of the calories from carbohydrate, 42% from fat (P/S ratio = 0.4), and 15% from protein in three meals daily, consuming 15% of calories at breakfast, 30% at lunch, and 55% at dinner. The diets provided a daily average of 550 mg of cholesterol and 5 g of crude fiber. The diets differed only in that 15% of the carbohydrate calories were provided by wheat starch alone, by 7.5% wheat starch and 7.5% fructose, or by fructose alone. The diets containing 7.5 and 15% fructose provided 50 and 100 g of fructose per day, respectively, based on an average intake of 2700 kcal. Each of the three diets was fed to the subjects in a crossover design for 5 weeks. Table 10 shows the effect of feeding the three different levels of fructose on fasting plasma triglycerides of hyperinsulinemic men and their normoinsulinemic controls. Triglyceride levels of the hyperinsulinemic men increased significantly as the level of fructose increased ($p < 0.05$). A similar pattern of increase in fasting plasma triglycerides of hyperinsulinemic men was observed as the sucrose content of the diet was increased from 5 to 18 to 33% of calories at the expense of wheat starch.[13] In contrast to the effect on hyperinsulinemic men, the level of fructose did not affect triglycerides in the normoinsulinemic controls. In a more recent study,[101] hyperinsulinemic men fed a diet typically American but low in copper (1 mg/day) showed significantly ($p < 0.05$) higher fasting plasma triglycerides when 20% of the calories was provided by fructose (317 mg %) than when 20% of the calories was provided by cornstarch (169 mg %). In contrast, normoinsulinemic men showed no difference in triglyceride levels when fed fructose (81 mg %) as compared to cornstarch (72 mg %). A number of experimental variables may explain the differences in the ability of similar amounts of fructose, fed for similar time periods to men exhibiting hypertriglyceridemia, to affect plasma triglycerides in these studies.[99,100] The effect of fructose was compared to the effect of either dextromaltose or starch. Dextromaltose is comprised of 56% maltose and 41% dextrins. The use of large

Table 10
FASTING PLASMA
TRIGLYCERIDES (mg %) OF
HYPER- AND
NORMOINSULINEMIC MEN FED
DIETS CONTAINING THREE
LEVELS OF FRUCTOSE FOR 5
WEEKS

	% fructose in diet		
Subject groups	**0**	**7.5**	**15**
Hyperinsulinemic men	102[a]	132[b]	163[c]
Normoinsulinemic men	86[a]	95[a]	92[a]

Note: Each value represents the mean of 60 values (5 weekly values for each of 12 men per group). Means not sharing a common superscript letter are significantly different from each other ($p < 0.05$).

Adapted from Hallfrisch, J., Reiser, S., and Prather, E. S., *Am. J. Clin. Nutr.*, 37, 740, 1983.

amounts of maltose as the comparison carbohydrate may have obscured a fructose-induced increase in blood triglycerides, since dissaccharides such as maltose have been shown to induce hepatic lipogenic enzyme activity in rats greater than that induced by their monosaccharide equivalents.[102] The pattern of dietary intake can influence blood triglyceride levels. The meals were provided in a nibbling pattern in one study,[99] and in more of a gorging pattern in the other.[100] The consumption of the same number of calories in large, in frequent meals (gorging) as compared to smaller, more frequent meals (nibbling) has been shown to increase blood triglyceride levels in humans.[103] The expression of sucrose-induced increases in blood triglycerides appears to be greater in diets fed in a gorging rather than a nibbling pattern.[66] In extrapolating from studies utilizing sucrose,[66] the magnitude of the increase of blood triglycerides by fructose would be expected to be greater when the dietary fat is saturated rather than unsaturated. Therefore, differences in the amount and type of dietary fat fed in these studies could have contributed to the apparently contradictory findings reported.

From these relatively few human studies employing a diversity of experimental designs, it appears possible to draw the following conclusions:

1. Fructose intake at levels usually consumed in this country does not appear to increase blood triglycerides significantly in the large population group with normal fasting triglyceride levels and unimpaired glucose tolerance.
2. Fructose intake by that population group with defects in carbohydrate metabolism producing elevated levels of blood triglycerides or insulin appears to increase blood triglyceride levels, as compared to the effect of other sources of dietary carbohydrate. These increases may occur even at relatively low levels of fructose intake.
3. The nature of the subject group studied and the experimental design employed appear to have similar effects on both sucrose- and fructose-induced increases in blood triglycerides, supporting the contention that the fructose moiety of sucrose is the responsible lipogenic agent.

2. Blood Cholesterol

Table 11 summarizes the effects of fructose feeding on blood cholesterol levels in humans,

Table 11

EFFECT OF FEEDING FRUCTOSE AS COMPARED TO OTHER CARBOHYDRATE SOURCES ON BLOOD CHOLESTEROL LEVELS IN HUMANS

Experimental conditions	Results	Ref.
Five young men (21—28 years), six young women (21—36 years), and three older women (58—68 years) were fed a fat-free diet containing 50 g of calcium caseinate, vitamins, and 7.5 g carbohydrate/kg body weight of 40% fructose, 60% cornstarch; 40% glucose, 60% cornstarch; and 40% fructose, 60% glucose each daily for 5 days	Significant ($p < 0.01$) decreases in serum cholesterol from pretest values in the older women after consuming all the fat-free diets; a strong trend toward decreases in serum cholesterol from pretest values in men after consuming all the fat-free diets; no change in serum cholesterol in young women due to dietary changes	88
Hypertriglyceridemic subjects were fed a diet containing an average of 18% protein, 5% fat, and 77% carbohydrate including 300 g per day of added fructose, glucose, or sucrose at the expense of starch for 2—4 weeks	Fructose significantly increased serum cholesterol in those subjects with a marked hypertriglyceridemia (above 300 mg %) in which fructose markedly increased serum triglycerides	91
Hypertriglyceridemic subjects were fed a diet containing 45% of calories as carbohydrate (of which 80% was from starch), 35% as fat (mainly saturated), and 20% as protein; fructose replaced 75—90 g of the starch during one of the 10- to 20-day feeding periods	The serum cholesterol concentration was not significantly influenced by the nature of the dietary carbohydrate (starch as compared to fructose)	93
Subjects consumed their self-selected diets using fructose ($n = 35$), sucrose ($n = 33$), or xylitol ($n = 48$) as the only sweetener for 2 years	No significant differences in serum cholesterol levels between dietary groups during the study	94
Men with diagnosed coronary artery disease were fed a diet containing 62% of the calories as carbohydrate, 22% as fat, and 16% as protein for 4 days, and provided an average of 182 mg of cholesterol daily; during one 4-day period the carbohydrate contained either 2 g fructose or glucose/kg body weight ($n = 14$ each)	A small but significant ($p < 0.01$) decrease in mean serum cholesterol was observed with both sugars after 4 days as compared to pretest levels	95
Four young women and men (20—32 years of age) were fed diets containing 35—49% of calories from carbohydrate, 35—45% from fat (P/S ratio = 0.5—1.5), and 12—20% from protein for 2 weeks in a crossover pattern; 33% of the carbohydrate calories were provided either by sucrose or fructose	No differences in total, LDL, or HDL cholesterol due to the nature of the sugar consumed	96
Subjects ($n = 11$) free of overt disease were fed a diet containing 55% of calories from carbohydrate (24% being provided by sucrose), 30% from fat (P/S ratio of about 1), and 15% from protein, and providing 300 mg of cholesterol daily for 3—4 days (base line period); fructose then replaced the sucrose and was fed for an additional 2 weeks	A significant ($p < 0.05$) decrease in both total and HDL cholesterol was seen during the fructose-feding period	97
Subjects ($n = 7$) with noninsulin-dependent diabetes were fed sucrose (base line) and then fructose using the same experimental conditions as those described for the previous study	No significant changes in either total or HDL cholesterol were observed after the fructose-feeding period	98
Subjects (12 normoinsulinemic men and 12 hyper-insulinemic men) were fed diets containing 43% of calories from carbohydrate (in which 0, 7.5, or 15% was provided by fructose at the expense of wheat starch), 42% from fat (P/S ratio = 0.4), and 15% from protein in a crossover pattern for 5 weeks each; the diet provided an average of 550 mg of cholesterol daily	Plasma total and LDL cholesterol levels were significantly ($p < 0.05$) greater after the subjects consumed the diet containing 7.5 or 15% fructose than after the diet containing no added fructose	100

from the relatively few studies reporting these effects.[88,91,93-98,100] In contrast to the effects on fasting blood triglycerides, the action of dietary carbohydrates on blood cholesterol is usually small. The nature and level of the dietary fat, especially dietary cholesterol, mainly determine levels of blood cholesterol. This can be illustrated by three of the studies listed in Table 11 in which the level of dietary fat and cholesterol either was severely reduced as compared to that expected from self-selected diets,[95,97] or was completely eliminated.[88] In each of these studies a significant decrease in blood cholesterol as compared to pretest levels was reported after feeding periods as short as 4 days and irrespective of the nature of the dietary carbohydrate. In two of the studies in which subjects with defects in carbohydrate metabolism were used,[91,100] the feeding of fructose as compared to starch produced significant increases in blood cholesterol levels. These increases, however, were small in comparison to the concurrent increases in endogenous blood triglycerides due to fructose feeding in these subjects. These increases in blood cholesterol may be attributed, at least in part, to the elevated levels of VLDL, containing 10 to 20% cholesterol,[104] produced in the liver in response to increased triglyceride synthesis. These results also suggest that conditions favoring fructose-induced increases in blood triglycerides (e.g., the use of diets containing high levels of saturated fat[105,106]) could contribute to the numerous environmental factors producing the high levels of blood cholesterol that are common in industrialized societies.

REFERENCES

1. **Olefsky, J. M., Farquhar, J. W., and Reaven, G. M.,** Reappraisal of the role of insulin in hypertriglyceridemia, *Am. J. Med.,* 57, 551, 1974.
2. **Reaven, G. M. and Bernstein, R. M.,** Effect of obesity on the relationship between very low density lipoprotein production rate and plasma triglyceride concentration in normal and hypertriglyceridemic subjects, *Metabolism,* 27, 1047, 1978.
3. **Carlson, L. A. and Bottiger, L. E.,** Ischemic heart disease in relation to fasting values of plasma triglycerides and cholesterol, *Lancet,* 1, 865, 1972.
4. **Tzagournis, M.,** Triglycerides in clinical medicine. A review, *Am. J. Clin. Nutr.,* 3, 1437, 1978.
5. **West, K. M., Ahuja, M. M. S., Bennett, P. H., Czyzyk, A., De Acosta, O. M., Fuller, J. H., Grab, B., Grabauskas, V., Jarrett, R. J., Kosaka, K., Keen, H., Krolewski, A. S., Miki, E., Schliack, V., Teuscher, A., Watkins, P. J., and Stober, J. A.,** The role of circulating glucose and triglyceride concentrations and their interactions with other "risk factors" as determinants of arterial disease in nine diabetic population samples from the WHO multinational study, *Diabetes Care,* 6, 361, 1983.
6. **Farquhar, J. W., Frank, A., Gross, R. C., and Reaven, G. M.,** Glucose, insulin and triglyceride response to high and low carbohydrate diets in man, *J. Clin. Invest.,* 45, 1648, 1966.
7. **Tzagournis, M., Chiles, R., Ryan, J. M., and Skillman, T. G.,** Interrelationships of hyperinsulinism and hypertriglyceridemia in young patients with coronary heart disease, *Circulation,* 38, 1156, 1968.
8. **Kuo, P. T. and Feng, L. Y.,** Study of serum insulin in patients with endogenous hypertriglyceridemia, *Metabolism,* 19, 372, 1970.
9. **Valek, J., Slabochová, Z., Grafnetter, D., and Kohout, M.,** The importance of stimulated immunoreactive-insulin levels to early results of dietary management in hyperlipoproteinemia, *Nutr. Metab.,* 17, 289, 1974.
10. **Bernstein, R. S., Grant, N., and Kipnis, D. M.,** Hyperinsulinemia and enlarged adipocytes in patients with endogenous hyperlipoproteinemia without obesity or diabetes mellitus, *Diabetes,* 24, 207, 1975.
11. **Julius, U., Schulze, J., Scholberg, K., Leonhardt, W., Hanefeld, M., Zschornack, M., and Haller, H.,** Influence of diet composition on insulin output, carbohydrate tolerance and lipid values in primary hypertriglyceridemia (HT), *Endokrinologie,* 71, 299, 1978.
12. **Bernstein, R. M., Davis, B. M., Olefsky, J. M., and Reaven, G. M.,** Hepatic responsiveness in patients with endogenous hypertriglyceridemia, *Diabetologia,* 14, 249, 1978.

13. **Reiser, S., Bickard, M. C., Hallfrisch, J., Michaelis, O. E., IV, and Prather, E. S.**, Blood lipids and their distribution in hyperinsulinemic subjects fed three different levels of sucrose, *J. Nutr.*, 111, 1045, 1981.

14. **Fuller, J. H., Shipley, M. J., Rose, G., Jarrett, R. J., and Keen, H.**, Coronary-heart-disease risk and impaired glucose tolerance, the Whitehall study, *Lancet*, 1, 1373, 1980.

15. **Norum, K. R.**, Some present concepts concerning diet and prevention of coronary heart disease, *Nutr. Metab.*, 22, 1, 1978.

16. **Witzum, J. and Schonfeld, G.**, High density lipoproteins, *Diabetes*, 28, 326, 1979.

17. **Miller, N. E., Forde, O. H., Thelle, D. S., and Mjøs, O. D.**, The Tromso heart study. High density lipoprotein and coronoary heart-disease: a prospective case-control study, *Lancet*, 1, 965, 1977.

18. **Hill, R., Baker, N., and Chaikoff, I. L.**, Altered metabolic patterns induced in the normal rat by feeding an adequate diet containing fructose as sole carbohydrate, *J. Biol. Chem.*, 209, 705, 1964.

19. **Zakim, D., Pardini, R. S., Herman, R. H., and Sauberlich, H. E.**, Mechanism for the differential effects of high carbohydrate diets on lipogenesis in rat liver, *Biochim. Biophys. Acta*, 144, 242, 1967.

20. **Bar-On, H. and Stein, Y.**, Effect of glucose and fructose administration on lipid metabolism in the rat, *J. Nutr.*, 94, 95, 1968.

21. **Pereira, J. N. and Jangaard, N. O.**, Different rates of glucose and fructose metabolism in rat liver tissue in vitro, *Metabolism*, 20, 392, 1971.

22. **Vessal, M., Choun, M. O., Bissell, M. J., and Bissell, D. M.**, Fructose utilization and altered cytochrome p-450 in cultured hepatocytes from adult rats, *Biochim. Biophys. Acta*, 633, 201, 1980.

23. **Topping, D. L. and Mayes, P. A.**, Comparative effects of fructose and glucose on the lipid and carbohydrate metabolism of perfused rat liver, *Br. J. Nutr.*, 36, 113, 1976.

24. **Baron, P., Griffaton, G., and Lowy, R.**, Metabolic inductions in the rat after an intraperitoneal injection of fructose and glucose, according to the nature of the dietary carbohydrate. II. Modifications after seven months of diet, *Enzyme*, 12, 481, 1971.

25. **Maruhama, Y. and Macdonald, I.**, Some changes in the triglyceride metabolism of rats on high fructose or glucose diets, *Metabolism*, 21, 835, 1972.

26. **Chevalier, M. M., Wiley, J. H., and Leveille, G. A.**, Effect of dietary fructose and fatty acid synthesis in adipose tissue and liver of the rat, *J. Nutr.*, 102, 337, 1972.

27. **Romsos, D. R. and Leveille, G. A.**, Effect of dietary fructose on in vitro and in vivo fatty acid synthesis in the rat, *Biochim. Biophys. Acta*, 360, 1, 1974.

28. **Tuovinen, C. G. R. and Bender, A. E.**, Some metabolic effects of prolonged feeding of starch, sucrose, fructose and carbohydrate-free diet in the rat, *Nutr. Metab.*, 19, 161, 1975.

29. **Blumenthal, M. D., Abraham, S., and Chaikoff, I. L.**, Dietary control of liver glucokinase activity in the normal rat, *Arch. Biochim. Biophys.*, 104, 215, 1964.

30. **Hornichter, R. D. and Brown, J.**, Relationship of glucose tolerance to hepatic glucokinase activity, *Diabetes*, 18, 257, 1969.

31. **Adelman, R. C., Spolter, P. D., and Weinhouse, S.**, Dietary and hormonal regulation of enzymes of fructose metabolism in rat liver, *J. Biol. Chem.*, 241, 5467, 1966.

32. **Nikkilä, E. A. and Ojala, K.**, Induction of hyperglyceridemia by fructose in the rat, *Life Sci.*, 4, 937, 1965.

33. **Cohen, A. M. and Teitelbaum, A.**, Effect of glucose, fructose and starch on lipogenesis in rats, *Life Sci.*, 7, 23, 1968.

34. **Sullivan, A. C., Miller, O. N., Wittman, J. S., III, and Hamilton, J. G.**, Factors influencing the in vivo and in vitro rates of lipogenesis of rat liver, *J. Nutr.*, 101, 265, 1971.

35. **Waddell, M. and Fallon, H. J.**, The effect of high-carbohydrate diets on liver triglyceride formation in the rat, *J. Clin. Invest.*, 52, 2725, 1973.

36. **Herzberg, G. R. and Rogerson, M.**, Dietary corn oil does not suppress the fructose induced increase in hepatic fatty acid synthesis, *Nutr. Res.*, 1, 73, 1981.

37. **Allen, R. J. L. and Leahy, J. S.**, Some effects of dietary dextrose, fructose, liquid glucose and sucrose in the adult male rat, *Br. J. Nutr.*, 20, 339, 1966.

38. **Bruckdorfer, K. R., Khan, I. H., and Yudkin, J.**, Fatty acid synthetase activity in the liver and adipose tissue of rats fed with various carbohydrates, *Biochem. J.*, 129, 439, 1972.

39. **Laube, H., Klör, H. U., Fussganger, R., and Pfeiffer, E. F.**, The effect of starch, sucrose, glucose and fructose on lipid metabolism in rats, *Nutr. Metab.*, 15, 273, 1973.

40. **Michaelis, O. E., IV and Szepesi, B.**, Effect of various sugars on hepatic glucose-6-phosphate dehydrogenase, malic enzyme and total liver lipid of the rat, *J. Nutr.*, 103, 697, 1973.

41. **Naismith, D. J. and Rana, I. A.**, Sucrose and hyperlipidaemia, *Nutr. Metab.*, 16, 285, 1974.

42. **Roggeveen, A. E., Geisler, R. W., Peary, D. E., and Hansen, R. J.**, Effects of diet on the activities of enzymes related to lipogenesis in rat liver and adipose tissue, *Proc. Soc. Exp. Biol. Med.*, 147, 467, 1974.

43. Vrána, A., Fábry, P., Slabochová, Z., and Kazdová, L., Effect of dietary fructose on free fatty acid release from adipose tissue and serum free fatty acid concentration in the rat, *Nutr. Metab.*, 17, 74, 1974.

44. Kritchevsky, D., Davidson, L. M., Shapiro, I. L., Kim, H. K., Kitagawa, M., Malhotra, S., Nair, P. P., Clarkson, T. B., Bersohn, I., and Winter, P. A. D., Lipid metabolism and experimental atherosclerosis in baboons: influence of cholesterol-free, semi-synthetic diets, *Am. J. Clin. Nutr.*, 27, 29, 1974.

45. Michaelis, O. E., IV, Nace, C. S., and Szepesi, B., Demonstration of a specific metabolic effect of dietary disaccharides in the rat, *J. Nutr.*, 105, 1186, 1975.

46. Waterman, R. A., Romsos, D. R., Tsai, A. C., Miller, E. R., and Leveille, G. A., Effects of dietary carbohydrate source on growth, plasma metabolites and lipogenesis in rats, pigs and chicks, *Proc. Soc. Exp. Biol. Med.*, 150, 220, 1975.

47. Sugawa-Katayama, Y. and Morita, N., Effects of a high fructose diet on lipogenic enzyme activities in some organs of rats fed ad libitum, *J. Nutr.*, 105, 1377, 1975.

48. Heller, G., Förster, H., and Fortmeyer, H. P., Influence of the various dietary carbohydrates on the concentration of metabolites and the activity of enzymes in the liver of diabetic rats, *Nutr. Metab.*, 21 (Suppl. 1), 177, 1977.

49. Kritchevsky, D., Davidson, L. M., Kim, H. K., Krendel, D. A., Malhotra, S., Mendelsohn, D., van der Watt, J. J., duPlessis, J. P., and Winter, P. A. D., Influence of type of carbohydrate on atherosclerosis in baboons fed semipurified diets plus 0.1% cholesterol, *Am. J. Clin. Nutr.*, 33, 1869, 1980.

50. Aoyama, Y., Yoshino, K., Yoshida, A., and Ashida, K., Triglyceride and cholesterol in serum and lipids in liver of rats fed either glucose or fructose from a diet and a drinking fluid, *Nutr. Rep. Int.*, 26, 1061, 1982.

51. Bird, M. I. and Williams, M. A., Triacylglycerol secretion in rats: effects of essential fatty acids and influence of dietary sucrose, glucose or fructose, *J. Nutr.*, 112, 2267, 1982.

52. Blakely, S. R., Hallfrisch, J., and Reiser, S., Long-term effects of moderate fructose feeding on lipogenic parameters in Wistar rats, *Nutr. Rep. Int.*, 25, 675, 1982.

53. Vrána, A., Raulin, J., Loriette, C., and Kazdová, L., Basal pyruvate dehydrogenase activity in the liver, adipose tissue and brain of rats with fructose-induced hypertriglyceridemia, *Nutr. Rep. Int.*, 28, 1437, 1983.

54. Reiser, S., Ferretti, R. J., Fields, M., and Smith, J. C., Jr., Role of dietary fructose in the enhancement of mortality and biochemical changes associated with copper deficiency in rats, *Am. J. Clin. Nutr.*, 38, 214, 1983.

55. Fields, M., Ferretti, R. J., Judge, J. M., Smith, J. C., and Reiser, S., Effects of different dietary carbohydrates on hepatic enzymes of copper-deficient rats, *Proc. Soc. Exp. Biol. Med.*, 178, 362, 1985.

56. Bruckdorfer, K. R., Kari-Kari, B. P. B., Khan, I. H., and Yudkin, J., Activity of lipogenic enzymes and plasma triglyceride levels in the rat and the chicken as determined by the nature of the dietary fat and carbohydrate, *Nutr. Metab.*, 14, 228, 1972.

57. Mavrias, D. A. and Mayer, R. J., Metabolism of fructose in the small intestine. II. The effect of fructose feeding on fructose transport and metabolism in guinea pig small intestine, *Biochim. Biophys. Acta*, 291, 538, 1973.

58. Leveille, G. A., Akinbami, T. K., and Ikediobi, C. O., Fructose adsorption and metabolism by the growing chick, *Proc. Soc. Exp. Biol. Med.*, 135, 483, 1970.

59. Mapungwana, S. M. and Davies, D. R., The effect of fructose on pyruvate kinase activity in isolated hepatocytes. Inhibition by allantoin and alanine, *Biochem. J.*, 208, 171, 1982.

60. Cohen, A. M., Teitelbaum, A., and Rosenmann, E., Diabetes induced by a high fructose diet, *Metabolism*, 26, 17, 1977.

61. Cohen, A. M., Genetically determined response to different ingested carbohydrates in the production of diabetes, *Horm. Metab. Res.*, 10, 86, 1978.

62. Sleder, J., Chen, Y.-D. I., Cully, M. D. and Reaven, G. M., Hyperinsulinemia in fructose-induced hypertriglyceridemia in the rat, *Metabolism*, 29, 303, 1980.

63. Zavaroni, I., Sander, S., Scott, S., and Reaven, G. M., Effect of fructose feeding on insulin secretion and insulin action in the rat, *Metabolism*, 29, 970, 1980.

64. Blakely, S. R., Hallfrisch, J., Reiser, S., and Prather, E. S., Long-term effects of moderate fructose feeding on glucose tolerance parameters in rats, *J. Nutr.*, 111, 307, 1981.

65. Tobey, T. A., Moden, C. E., Zavaroni, I., and Reaven, G. M., Mechanism of insulin resistance in fructose-fed rats, *Metabolism*, 31, 608, 1982.

66. Reiser, S., Physiological differences between starches and sugars, in *Medical Applications of Clinical Nutrition*, Bland, J. and Shealy, N., Eds., Keats Publishing, New Canaan, Conn., 1983, chap. 6.

67. Cook, G. C., Absorption and metabolism of D (-)fructose in man, *Am. J. Clin. Nutr.*, 24, 1302, 1971.

68. Mavrias, D. A. and Mayer, R. J., Metabolism of fructose in the small intestine. I. The effect of fructose feeding on fructose transport and metabolism in rat small intestine, *Biochim. Biophys. Acta*, 291, 531, 1973.

69. **Heinz, F., Lamprecht, W., and Kirsch, J.,** Enzymes of fructose metabolism in human liver, *J. Clin. Invest.,* 47, 1826, 1968.

70. **Zakim, D., Herman, R. H., and Gordon, W. C., Jr.,** The conversion of glucose and fructose to fatty acids in the human liver, *Biochem. Med.,* 2, 427, 1969.

71. **Macdonald, I.,** Ingested glucose and fructose in serum lipids in healthy men and after myocardial infarction, *Am. J. Clin. Nutr.,* 21, 1366, 1968.

72. **Maruhama, Y.,** Conversion of ingested carbohydrate-^{14}C into glycerol and fatty acids of serum triglyceride in patients with myocardial infarction, *Metabolism,* 19, 1085, 1970.

73. **Wolfe, B. M. and Ahuja, S. P.,** Effects of intravenously administered fructose and glucose on splanchnic secretion of plasma triglycerides in hypertriglyceridemic men, *Metabolism,* 26, 963, 1977.

74. **Chevalier, M., Wiley, J. H., and Leveille, G. A.,** The age-dependent response of serum triglycerides to dietary fructose, *Proc. Soc. Exp. Biol. Med.,* 139, 220, 1972.

75. **Vrána, A., Fábry, P., and Kazdová, L.,** Effect of dietary fructose on fatty acid synthesis in adipose tissue and on triglyceride concentration in blood in the rat, *Nutr. Metab.,* 15, 305, 1973.

76. **Thenen, S. W. and Mayer, J.,** Effect of fructose and other dietary carbohydrates on plasma glucose, insulin, and lipids in genetically obese (Ob/Ob) mice (39569), *Proc. Soc. Exp. Biol. Med.,* 153, 464, 1976.

77. **Vrána, A., Fábry, P., and Kazdová, L.,** Effects of dietary fructose on serum triglyceride concentrations in the rat, *Nutr. Rep. Int.,* 14, 593, 1976.

78. **Vrána, A., Poledne, R., Fábry, P., and Kazdová, L.,** Palmitate and glucose oxidation by diaphragm of rats with fructose-induced hypertriglyceridemia, *Metabolism,* 27, 885, 1978.

79. **Kang, S. S., Bruckdorfer, K. R., and Yudkin, J.,** Influence of different dietary carbohydrates on liver and plasma constituents in rats adapted to meal feeding, *Nutr. Metab.,* 23, 301, 1979.

80. **Merkens, L. S., Tepperman, H. M., and Tepperman, J.,** Effects of short-term dietary glucose and fructose on rat serum triglyceride concentration, *J. Nutr.,* 110, 982, 1980.

81. **Aoyama, Y., Hattori, Y., Yoshida, A., and Ashida, K.,** Lipoprotein lipase activity of adipose tissue in rats fed ad libitum and refed a diet containing glucose or fructose, *Nutr. Rep. Int.,* 24, 839, 1981.

82. **Kannan, R., Baker, N., and Bruckdorfer, K. R.,** Secretion and turnover of very low density liproprotein triacylglycerols in rats fed chronically diets rich in glucose and fructose, *J. Nutr.,* 111, 1216, 1981.

83. **Høstmark, A. T., Spydevold, Ø., Lystad, E., and Eilertsen, E.,** Plasma lipoproteins in rats fed starch, sucrose, glucose or fructose, *Nutr. Rep. Int.,* 25, 161, 1982.

84. **Zavaroni, I., Chen, Y.-D. I., and Reaven, G. M.,** Studies of the mechanism of fructose-induced hypertriglyceridemia in the rat, *Metabolism,* 31, 1077, 1982.

85. **Høstmark, A. T., Spydevold, Ø., Lystad, E., and Hang, A.,** Apoproteins in lipoproteins of d < 1.063 g/mℓ from rats fed glucose and fructose diets, *Nutr. Rep. Int.,* 29, 1361, 1984.

86. **Hämäläinen, M. M. and Mäkinen, K. K.,** Metabolism of glucose, fructose and xylitol in normal and streptozotocin-diabetic rats, *J. Nutr.,* 112, 1369, 1982.

87. **Westring, M. E. and Potter, N. N.,** Rat feeding studies using unheated and heated mixtures of casein with fructose, glucose/fructose, and sucrose, *Nutr. Rep. Int.,* 29, 371, 1984.

88. **Macdonald, I.,** Influence of fructose and glucose on serum lipid levels in men and pre- and postmenopausal women, *Am. J. Clin. Nutr.,* 18, 369, 1966.

89. **Macdonald, I.,** The lipid response of postmenopausal women to dietary carbohydrates, *Am. J. Clin. Nutr.,* 18, 86, 1966.

90. **Macdonald, I.,** The lipid response of young women to dietary carbohydrates, *Am. J. Clin. Nutr.,* 16, 458, 1965.

91. **Kaufmann, N. A., Poznanski, R., Blondheim, S. H., and Stein, Y.,** Effect of fructose, glucose, sucrose and starch on serum lipids in carbohydrate induced hypertriglyceridemia and in normal subjects, *Isr. J. Med. Sci.,* 2, 715, 1966.

92. **Kaufmann, N. A., Kapitulnik, J., and Blondheim, S. H.,** Studies in carbohydrate-induced hypertriglyceridemia. Aspects of fructose metabolism, *Isr. J. Med. Sci.,* 6, 80, 1970.

93. **Nikkilä, E. A.,** Influence of dietary fructose and sucrose on serum triglycerides in hypertriglyceridemia and diabetes, in *Sugars in Nutrition,* Sipple, H. L. and McNutt, K. W., Eds., Academic Press, Orlando, Fla., 1974, chap. 26.

94. **Huttunen, J. K., Mäkinen, K. K., and Scheinin, A.,** Turku sugar studies. XI. Effects of sucrose, fructose and xylitol diets on glucose, lipid and urate metabolism, *Acta Odont. Scand.,* 33(Suppl.70), 239, 1975.

95. **Palumbo, P. J., Briones, E. R., Nelson, R. A., and Kottke, B. A.,** Sucrose sensitivity of patients with coronary-artery disease, *Am. J. Clin. Nutr.,* 30, 394, 1977.

96. **Bossetti, B. M., Kocher, L. M., Moranz, J. F., and Falko, J. M.,** The effects of physiologic amounts of simple sugars on lipoprotein, glucose, and insulin levels in normal subjects, *Diabetes Care,* 7, 309, 1984.

97. **Crapo, P. A. and Kolterman, O. G.,** The metabolic effects of 2-week fructose feeding in normal subjects, *Am. J. Clin. Nutr.,* 39, 525, 1984.

98. **Crapo, P. A., Kolterman, O. G., and Henry, R. R.,** The metabolic consequence of 2-week fructose feeding in diabetic subjects, *Diabetes Care,* 9, 111, 1986.
99. **Turner, J. L., Bierman, E. L., Brunzell, J. D., and Chait, A.,** Effect of dietary fructose on triglyceride transport and glucoregulatory hormones in hypertriglyceridemic men, *Am. J. Clin. Nutr.,* 32, 1043, 1979.
100. **Hallfrisch, J., Reiser, S., and Prather, E. S.,** Blood lipid distribution of hyperinsulinemic men consuming three levels of fructose, *Am. J. Clin. Nutr.,* 37, 740, 1983.
101. **Reiser, S., Smith, J. C., Jr., Scholfield, D. J., Powell, A. S., Yang, C.-Y., and Mertz, W.,** Indices of copper status and lipogenesis in humans consuming a typical American diet containing either fructose or starch, *Fed. Proc., Fed. Am., Soc. Exp. Biol.,* 44, 540, 1985.
102. **Michaelis, O. E., IV and Szepesi, B.,** Specificity of the disaccharide effect in the rat, *Nutr. Metab.,* 21, 329, 1977.
103. **Young, C. M., Hutter, L. F., Scanlan, S. S., Rand, C. E., Lutwak, L., and Simko, V.,** Metabolic effects of meal frequency on normal young men, *J. Am. Diet. Assoc.,* 61, 391, 1972.
104. **Fredrickson, D. S., Levy, R. I., and Lees, R. S.,** Fat transport in lipoproteins — an integrated approach to mechanisms and disorders, *N. Engl. J. Med.,* 276, 94, 1967.
105. **Antar, M. A., Little, J. A., Lucas, C., Buckley, G. C., and Csima, A.,** Interrelationship between the kinds of dietary carbohydrate and fat in hyperlipoproteinemic patients. III. Synergistic effect of sucrose and animal fat on serum lipids, *Atherosclerosis,* 11, 191, 1970.
106. **Mann, J. I., Watermeyer, G. S., Manning, E. B., Randles, J., and Truswell, A. S.,** Effects on serum lipids of different dietary fats associated with a high sucrose diet, *Clin. Sci.,* 44, 601, 1973.

Chapter 8

URIC ACID AND LACTIC ACID

I. INTRODUCTION

Uric acid and lactic acid are compounds produced in the course of normal metabolic processes in mammalian tissues. In this chapter the evidence for the association between high levels of blood uric acid and various diseases will be presented; the acute and longer-range effects on blood uric acid and lactic acid observed after the administration of fructose or the fructose-containing disaccharide, sucrose, to humans will be described; and the mechanisms proposed to explain some of these effects discussed.

II. URIC ACID

Uric acid is the end product of the catabolism of purine nucleotides in humans and higher primates. The formation of uric acid occurs primarily in the liver. It is a normal component of human blood; the confirmed normal range of blood uric acid is 3.8 to 4.5 mg %, but considerable variations about this confirmed range have been reported.[1] The increase in the blood levels of uric acid above the normal range has been associated directly or indirectly with disease states. Increased levels of uric acid are found in humans with gout and are associated with an increased risk of heart disease.

A. Association with Disease States and Metabolic Abnormalities
1. Heart Disease

The incidence of elevated blood uric acid in patients with coronary heart disease has been found to be alarmingly high. Serum uric acid values of 6.0 mg % for men and 5.0 mg % for women are generally considered to represent the upper limits of normal. Approximately 6 to 10% of the general population have levels in the abnormal range.[2] As early as 1951, it was reported that hyperuricemia was present four times as frequently in men who had myocardial infarction before the age of 40 than in a control group.[3] This finding was soon confirmed by Kohn and Prozan,[4] who reported that in a group of 50 patients with a myocardial infarction, 57% of the men and 67% of the women had respective serum uric acid levels above 6 and 5 mg %. A higher frequency of coronary events has also been observed in patients with gout.[2] Using levels above 5.3 and 4.3 mg % as abnormal for men and women, respectively, it was found that 83% of the men and 63% of the women from a group of 280 subjects with diagnosed myocardial infarctions had serum uric acid levels greater than control means for their sex.[5] In a similar type of study of the level of serum uric acid in patients suffering from a myocardial infarction and patients with ischemic heart disease were compared with levels in a group of normal subjects.[6] The healthy individuals averaged 4.53 and 3.36 mg % serum uric acid for the males and females, respectively. In contrast, the serum uric acid values were 6.89 and 4.87 mg % for the males and females in the myocardial infarction group and an overall average of 6.01 mg % in the ischemia group. The authors concluded that hyperuricemia was one of the clinical abnormalities associated with myocardial infarction.

A few studies have attempted to determine whether high levels of uric acid are a predictor of cardiovascular disease and its resultant mortality. High levels of serum uric acid were reported to be significantly associated with an increased risk of coronary heart disease.[7] A group consisting of 8341 male subjects with elevated serum uric acid levels and established myocardial disease was studied as part of a long-range (at least 3 years) double-blind trial, the main objective of which was to determine the effect of lipid-lowering drugs on subsequent death from cardiovascular disease.[8] The 2789 men reported on represented the placebo group, one third of all subjects studied. Serum uric acid levels above 7 mg % were observed in 44% of these subjects. Results of univariate analysis showed that total mortality rate was higher in subjects with higher serum uric acid and lower in those with lower serum uric acid values. However, the relationship in this analysis between serum uric acid and cardiovascular mortality was not found to be significant after adjustment for diuretic use (thiazide diuretics frequently prescribed for hypertension have a well-known side effect of hyperuricemia). It was therefore concluded that raised serum uric acid had little independent predictive power for cardiovascular mortality. However, this conclusion was contradicted on the basis of a study with 2758 male and 2011 female hyperuricemic subjects singled out by a multiphasic health screening program in Finland.[9] After a median follow-up time of 6 years, the 4769 hyperuricemic subjects were found to have higher cardiovascular death rates than normouricemics. When adjustments were made for use of diuretics, multivariate analysis revealed that the hyperuricemics still had a significantly greater mortality rate, suggesting that serum uric acid can have an independent contribution in prediction of cardiovascular mortality.

The reason for the high incidence of hyperuricemia in patients with coronary vascular disease is unknown. It has been suggested that the observed increase in platelet turnover rate and decrease in platelet survival time which occur in hyperuricemia could result in increased platelet adhesiveness.[5] This may ease the ability of platelets to adhere to irregular or injured vessel walls and agglutinate. Rupture of platelets would cause release of 5-hydroxytryptamine which in turn causes vasoconstriction and, conceivably, intravascular thrombosis.

2. Blood Lipids

A positive association between the levels of serum uric acid and of blood triglycerides and/or cholesterol has been reported in a number of studies. These findings may explain in part the relationship between high levels of serum uric acid and the incidence of heart disease described in the previous section.

a. Cholesterol

In a study in which the changes in serum uric acid and serum cholesterol were monitored for 6 to 10 weeks in 14 patients with coronary vascular disease, the serum values were found to be positively correlated.[10] Not only did the serum uric acid and cholesterol parallel each other in direction but also in magnitude. A high incidence of hyperuricemia in hypertensive patients with hypercholesterolemia also has been observed.[11] Likewise, it was found that the incidence of both hyperuricemia and hypercholesterolemia was abnormally high in patients with coronary vascular disease.[4] A positive correlation between serum cholesterol and uric acid also has been observed in patients with adult-onset diabetes.[12]

In contrast to these findings, no significant correlation between serum uric acid and serum cholesterol was found in 282 patients with peripheral vascular disease.[13] Furthermore, it was reported that correlation analysis of previously published data[4,10] revealed no significant relationship between serum uric acid and serum cholesterol. A study of 185 members of 12 Danish families with familial hypercholesterolemia also failed to demonstrate a significant correlation or regression coefficient between serum uric acid and serum cholesterol.[14] Likewise, other researchers have failed to find a significant correlation between hyperuricemia and hypercholesterolemia.[2,9,15]

b. Triglycerides

It appears that serum uric acid is more closely correlated with blood triglyceride levels than with blood cholesterol.[7,13,16-19] The basis for this conclusion can be exemplified by the results from a study in which blood lipid and uric acid interrelationships were determined in 100 patients with hyperlipemia and in 50 normal controls.[18] Mean serum uric acid levels of 4.1 mg % for the control group and 4.4 mg % for hypercholesterolemic subjects were not significantly different. However, the group with elevations of both cholesterol and triglycerides had a mean serum uric acid of 7.0 mg %. This was significantly greater than the mean of either of the other two groups. Furthermore, the correlation coefficient between serum triglycerides and serum uric acid was 0.64, but only −0.07 between serum cholesterol and serum uric acid. It was also reported that of 25 gouty patients, 21 (84%) had abnormally high triglyceride levels, the mean of 217 mg % being significantly elevated above that of the control group. However, only 20% of the patients had elevated serum cholesterol levels. It was therefore concluded that the prior assumed relationship between serum uric acid and serum cholesterol was actually a relationship between serum uric acid and serum triglycerides.

The metabolic basis for the positive association between blood uric acid levels and triglyceride levels has not been resolved. It does not appear that hyperuricemia by itself produces increased levels of blood triglycerides.[20] Conversely, hypertriglyceridemia does not appear to be the cause of hyperuricemia.[19] It appears more likely that the concurrently elevated levels of blood uric acid and triglycerides are results of a metabolic disorder which causes the increase of both of these metabolites. For example, both hyperuricemia and hypertriglyceridemia are observed in two inborn errors of carbohydrate metabolism, resulting from a deficiency of glucose-6-phosphatase (glycogen storage disease or von Gierke's disease)[21-23] and of fructose-1,6-diphosphatase.[24]

3. Hypertension and Stroke

A relationship appears to exist between elevated levels of serum uric acid and hypertension. In a group of men with untreated hypertension, it was found that 25% had serum uric acid levels greater than 7.2 mg %.[25] Although some of the group had a history of alcohol consumption, the elevated serum uric acid was not totally attributable to the effects of alcohol. The 6.5% incidence of hypertension in 124 asymptomatic hyperuricemic subjects followed for 50 weeks was reported to be higher than that observed in a group of 224 normouricemic subjects.[26] In a survey of 1000 gouty subjects, some of whom were taking diuretics, 37% had arterial hypertension.[27] High levels of serum uric acid also have been observed in 250 patients with hypertension.[28]

Studies have shown that while the incidence of hyperuricemia is greater among hypertensives taking diuretics, the occurrence of hyperuricemia in untreated hypertensives is also higher than that of the population at large. It was reported that of 470 hypertensives on drug treatment, 54% were hyperuricemic. The incidence of hyperuricemia among untreated hypertensives was 27%, which is still two to four times higher than that of the general population.[29] Renal abnormality produced by the hypertension, with consequent decreased serum uric acid clearance, was proposed as the mechanism for the association between hypertension and hyperuricemia. A high incidence of hyperuricemia was found in another group of 119 hypertensives not taking diuretics.[11] In this study, 38% of the subjects had serum uric acid levels greater than 6.4 and 5.9 mg % for men and women, respectively. Furthermore, it was found that the incidence of hyperuricemia was independent of the type of hypertension, suggesting that hyperuricemia in hypertensive disease is related neither to genetic factors operative in essential hypertension nor to the etiology of underlying renal disease in hypertension. The finding of a marked increase in the incidence of hyperuricemia in patients with uremia and in those taking hypertensive drugs which work at the kidney level led to a conclusion favoring decreased renal excretion of uric acid as the mechanism for the association between hypertension and hyperuricemia. This mechanism has been questioned on the basis of results from a more recent study.[30] Nine hypertensive subjects and eight controls, all with normal fasting serum uric acid levels, were given 0.3 g fructose per kilogram body weight by i.v. infusion. Serum uric acid levels in response to the fructose were significantly greater in the hypertensives than in the controls. However, no significant difference in uric acid excretion was found between the two groups. Thus, a hypertensive-induced renal abnormality resulting in decreased uric acid excretion could not be the cause of the high incidence of hyperuricemia in hypertensive disease. It was concluded that the hyperuricemia resulted from increased catabolism of preformed purines induced by the fructose load, and proposed that hypertensives have a greater purine pool than normotensives.

Patients suffering from cerebrovascular disease were reported to have much higher serum uric acid levels than a group of control subjects.[31] Using values of 7 and 6 mg % as abnormal levels for males and females, the authors reported that 25% of the acute stroke patients were hyperuricemic. These findings were confirmed in another study in which it was noted that 30% of stroke victims were hyperuricemic.[15] Furthermore, it has been reported that the hyperuricemia seen in ischemic thrombotic cerebrovascular disease was not a transitory or secondary phenomenon,[15,31] indicating that hyperuricemia may be a pathogenic factor.

4. Gout

With an increased concentration of serum uric acid, the probability of the development of gout is increased.[2,32] Although age-dependent, the incidence of gout at a serum uric acid concentration of 9 mg % has been reported to be 90% in males.[2] It has been reported that 13.2% of a group of routinely screened hospitalized men were hyperuricemic.[33] Of 200 of these patients who were examined in detail, 12% were diagnosed as having gout resulting from the hyperuricemia. A method of diagnosing gout has been developed, using a 50-g fructose infusion.[34] Serum uric acid concentrations at 0 and 120 min were used to formulate the diagnosis and monitor the effect of treatment.

5. Glucose Tolerance

Several epidemiological studies have shown no evidence for an association between blood glucose and serum uric acid levels;[35-37] indeed, high levels of blood glucose appear to lower serum uric acid levels.[7,38-40] In an epidemiological survey of 8000 men of Japanese ancestry, diabetes was found to be negatively associated with serum uric acid.[7] Similar results were observed in three diabetes surveys taken at two yearly intervals.[40] Uric acid levels and the response to a glucose tolerance test were determined for 10,000 men. Four groups were compared: (1) prediabetics, represented by those who did not have diabetes initially but had developed diabetes by a later survey; (2) an abnormal glucose tolerance test group; (3) diabetics; and (4) those who remained normal throughout the time span of the surveys. The prediabetics had higher serum uric acid levels than the normal group. These men subsequently developed diabetes within 5 years or less. Those with an abnormal glucose tolerance test had serum uric acid levels slightly higher than normal. The diabetic group responded with much lower than normal values; it was also observed that the longer the duration of the diabetes, the lower the uric acid levels. It was hypothesized that the hyperglycemic and glycosuric states that existed in diabetes could have inhibited the reabsorption of uric acid by glucose in the proximal tubules, thus accounting for the lower serum uric acid levels seen in diabetics.

B. Effect of Fructose

1. Intravenous Administration

In studies in which fructose has been administered intravenously to human subjects in amounts of 0.5 g/kg body weight and above, rapid and significant increases in blood uric acid levels have usually resulted. The infusion of a 20% fructose solution to attain a level of 0.5 g/kg body weight in normal children and children with hereditary fructose intolerance

Table 1
SERUM URIC ACID LEVELS BEFORE AND 90 MIN
AFTER THE START OF AN INFUSION PROVIDING
THE INDICATED LEVELS OF FRUCTOSE IN
HEALTHY MEN

Rate of infusion (g/kg body weight per hour)	n	Serum uric acid levels (mg %)	
		0 min	90 min
0.5	5	5.2 ± 0.50	5.2 ± 0.60
1.0	8	4.6 ± 0.14	5.5 ± 0.70
			$p < 0.10$
1.5	8	5.0 ± 0.69	6.4 ± 0.57
			$p < 0.025$

Note: Each value represents the mean ± S.D. from the indicated number of subjects.

Adapted from Heuckenkamp, P.-U. and Zollner, N., *Lancet,* 1, 808, 1971.

produced both hyperuricemia and hyperuricosuria.[41] The serum uric acid levels were still elevated 5 hr after the infusion. Fructose-induced hyperuricemia was reported after 73 normal adults were given 50 g of fructose intravenously over a 30-min period (0.7 g/kg body weight for a subject weighing 70 kg).[42] In a study in which both healthy subjects and those with compensated cirrhosis of the liver were infused with fructose at the rate of 4 g/min for 6 min, a significant increase in serum uric acid levels occurred 10 min after completion of the infusions.[43] The rise in serum uric acid levels remained significant for as long as 40 min postinfusion.

The increase in serum uric acid produced by i.v. fructose administration appears to be specific for fructose and not shared by other common dietary monosaccharides. This specificity was demonstrated when eight healthy men were infused with 0.5 g/kg body weight over 15 min as a 20% solution of fructose, glucose, or galactose and blood uric acid levels were determined intermittently for 3 hr.[44] Fructose infusion produced a 30% increase in peak (30 min) plasma uric acid levels. In contrast, glucose and galactose did not increase plasma uric acid above control levels.

The increase in blood uric acid produced by fructose infusion appears to be dependent on the amount of fructose infused. This dose dependence may partially explain the failure of fructose infusion at levels around 0.5 g/kg body weight to increase serum uric acid in some studies using healthy subjects. For example, no changes in serum uric acid levels were found after ten healthy males were infused with 0.5 g fructose per kilogram body weight per hour.[45] Similarly, infusions of 20 and 50 g fructose per hour (0.3 and 0.7 g/kg body weight per hour, respectively, for a 70-kg man) to ten healthy men produced no significant increases in serum uric acid levels.[46] The dose-dependence of fructose-induced hyperuricemia was best illustrated by a study of 21 healthy men who received varying amounts of a 10% fructose solution intravenously over a 6-hr period.[47] One group received 0.5, another 1.0, and a third 1.5 g/kg body weight per hour. The results are shown in Table 1. Subjects who were infused with 0.5 g fructose per kilogram body weight per hour showed no significant change in serum uric acid from preinfusion levels. Fructose administered at a rate of 1.0 g/kg body weight per hour caused a 20% rise in serum uric acid ($p < 0.1$). A fructose load of 1.5 g/kg body weight per hour produced a significant 28% increase ($p < 0.025$) in uric acid, with one subject (data not included) doubling his preinfusion level. These results are consistent with those reported by Förster[48,49] and Förster and Hoos.[50] A large and significant

Table 2

**SERUM URIC ACID LEVELS BEFORE AND 30 TO 180
MIN AFTER CONTROLS AND NONKETOTIC
DIABETICS WERE INFUSED WITH FRUCTOSE TO
ACHIEVE A LEVEL OF 0.5 g/kg IDEAL BODY WEIGHT**

Time after infusion (min)	Serum uric acid levels (mg %)	
	Control (n = 10)	Nonketotic diabetics (n = 20)
0 (fasting)	3.7 ± 1.3	3.9 ± 1.3
30	4.0 ± 1.8	4.9 ± 1.8[a]
60	4.9 ± 1.0[a]	5.7 ± 1.5[a]
90	5.2 ± 1.8[a]	6.6 ± 1.4[a]
120	4.7 ± 2.0[a]	5.7 ± 2.0[a]
180	3.6 ± 1.2	4.7 ± 1.8[a]

Note: Each value represents the mean ± S.D. from the number of subjects shown in parentheses.

[a]   Significantly higher than fasting level.

Adapted from El-Ebrashy, W., Shaheen, M. H., Wasfi, A. A., and El-Dana-soury, M., *J. Egypt. Med. Assoc.*, 57, 406, 1974.

Table 3

**PLASMA URIC ACID RESPONSES BEFORE AND 15 TO
120 MIN AFTER HYPER- AND NORMOTENSIVE
SUBJECTS WERE GIVEN A FRUCTOSE LOAD OF 0.3 g/
kg BODY WEIGHT INTRAVENOUSLY IN ABOUT 10 MIN**

Time after fructose load (min)	Plasma uric acid responses (mg %)	
	Hypertensive (n = 9)	Normotensive (n = 8)
0	4.5 ± 0.3	4.2 ± 0.3
15	6.3 ± 0.5[a]	5.0 ± 0.3
30	6.2 ± 0.4[a]	4.8 ± 0.3
60	6.1 ± 0.5[a]	4.8 ± 0.3
120	5.5 ± 0.4	4.9 ± 0.3

Note: Each value is the mean ± SEM from the number of subjects shown in parentheses.

[a]   Significantly greater than comparable normotensive value ($p < 0.05$) by un-paired student's *t*-test.

Adapted from Fiaschi, E., Baggio, B., Favaro, S., Antonello, A., Camerin, E., Todesco, S., and Borsatti, A., *Metabolism*, 26, 1219, 1977.

increase in serum uric acid of adults following administration of 1.5 g fructose per kilogram body weight was observed. An increase of lesser magnitude was found after fructose infusion of 1 g/kg body weight. However, very little effect on serum uric acid was observed after infusion of 0.25 g fructose per kilogram body weight per hour. Fructose-induced hyperur-icemia has not been demonstrated in children at a dose lower than 0.5 g/kg body weight.[41] Consistent with this finding, a dose of 0.25 g fructose per kilogram body weight per hour was reported to have very little effect on serum uric acid levels in children.[51]

The magnitude of the fructose-induced increase in serum uric acid is also dependent on the state of health of the subjects tested. For example, patients with gout appear to respond more to fructose than do healthy subjects, and fructose-induced increases in blood uric acid levels in patients with gout appear to occur at lower fructose levels than in normal subjects. Significant increases in plasma uric acid ranging from 20 to 40% above control values were observed in four men with gout 30 to 180 min after being infused for 10 min with a volume of a 20% solution of fructose calculated to achieve the comparatively low level of 0.5 g/kg body weight.[52] Similarly, the infusion of fructose to achieve a level of about 0.6 g/kg body weight in four men with gout produced a mean increase in plasma uric acid of 22% above base line levels.[53] Uric acid levels in diabetics as compared to normal subjects also appear to be increased more readily by fructose infusion. Table 2 presents the serum uric acid levels before and 30 to 180 min after controls and nonketotic diabetics were infused with a 40% solution of fructose calculated to achieve a dose level of 0.5 g/kg ideal body weight in 4 min.[54] The diabetic subjects exhibited an earlier rise, a greater percentage increase above fasting, and a slower decrease in serum uric acid in response to fructose than did the controls. Since elevated levels of uric acid and gout[55] as well as impaired glucose tolerance are associated with type IV or carbohydrate-induced hyperlipoproteinemia, these results suggest that subjects with this genetic predisposition would be more susceptible to fructose-induced increases in blood uric acid than would normal individuals.

Subjects with hypertension also appear to be more responsive to fructose-induced increases in plasma uric acid than normotensive individuals. Table 3 presents the plasma uric acid responses before and 15 to 120 min after nine hypertensive subjects and eight controls were infused with a fructose load of 0.3 g/kg body weight in about 10 min.[30] Although increases in plasma uric acid due to fructose were observed in both subject groups, the hypertensives had significantly higher plasma uric acid levels than the normotensives after 15, 30, and 60 min. These differences could not be accounted for by pretest differences in plasma uric acid between the two subject groups.

2. Short-Term Oral Response

Several studies have shown that a single oral load of fructose can produce increases in blood uric acid levels. A significant increase in serum uric acid was observed 60 min after an oral load of 0.5 g/kg body weight in patients with gout.[56] The effect of the oral intake of 1 g/kg body weight of fructose on serum uric acid levels in six healthy subjects, six gouty subjects, and five children of gouty parents has been reported.[57] Fructose caused a rise in serum uric acid in the healthy subjects, but a more obvious and delayed increase was observed in the gouty subjects and their progeny. Two of the gouty subjects consumed a low-fructose diet for 1 year. At the end of this period, a fructose load of 1 g/kg body weight elicited a uric acid response similar to that of the healthy subjects. Conversely, when these subjects consumed for 3 weeks a diet rich in sweets and fruits, the uric acid response to a fructose load was the same as seen initially. Although this study used a small number of subjects, it indicated not only that dietary fructose could alter the uric acid response, but also that some individuals were genetically predisposed toward this abnormal response.

The ability of a carbohydrate to increase blood uric acid levels following oral intake appears to be dependent on the presence of the fructose moiety. The effect on serum uric acid levels before and 15 to 90 min after the ingestion of 1 g/kg body weight of glucose, sucrose, or fructose by nine healthy men is shown in Table 4.[58] Despite the differences in the levels of serum uric acid in the subjects within their respective dietary groups, it is apparent that uric acid levels increased only in those subjects receiving the fructose. As with the infusion studies, the effect of oral fructose on serum uric acid appears to be dose-dependent. The intake of fructose at levels of 0.5 g/kg body weight (the sucrose group) failed to increase serum uric acid in these healthy subjects. In a similar type of study,[59] 23

Table 4
SERUM URIC ACID LEVELS
BEFORE AND 15 TO 90 MIN
AFTER THE INGESTION OF 1 g/
kg BODY WEIGHT OF GLUCOSE,
SUCROSE, OR FRUCTOSE BY
HEALTHY MEN

Time (min)	Serum uric acid levels (mg %)		
	Glucose	Sucrose	Fructose
0	5.4 ± 0.3	6.1 ± 0.4	7.8 ± 0.6
15	6.0 ± 0.5	6.1 ± 0.4	8.0 ± 0.6
30	5.8 ± 0.3	6.3 ± 0.4	9.0 ± 0.6
60	5.5 ± 0.5	6.4 ± 0.4	8.8 ± 0.5
90	5.3 ± 0.4	6.3 ± 0.5	8.5 ± 0.6

Note: Each value represents the mean ± SEM from nine men.

Adapted from Macdonald, I., Keyser, A., and Pacy, D., *Am. J. Clin. Nutr.*, 31, 1305, 1978.

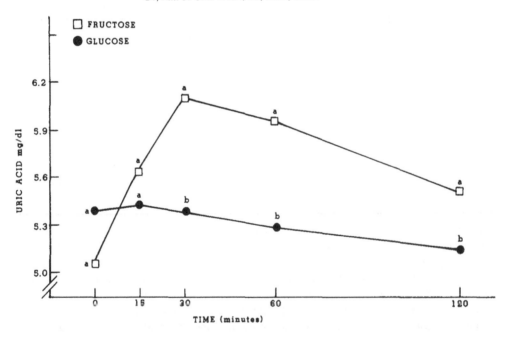

FIGURE 1. Plasma uric acid response after 23 normal subjects consumed a drink containing 1 g/kg body weight of either glucose or fructose. Values at each time point not sharing a common superscript letter are significantly different ($p < 0.05$) as determined by Duncan's multiple range test. (Adapted from Reiser, S., Scholfield, D., Trout, D., Wilson, A., and Aparicio, P., *Nutr. Rep. Int.*, 30, 151, 1984.)

healthy subjects were given 1 g/kg body weight of either a glucose or fructose drink in a crossover pattern separated by at least 3 weeks. Blood was drawn before and 15, 30, 60, and 120 min after the drink and plasma was analyzed for uric acid (Figure 1). Plasma uric acid increased significantly above fasting levels at all times except 120 min only after intake of the fructose-containing drink. Uric acid levels were significantly higher 30, 60, and 120 min after intake of the fructose- as compared to the glucose-containing drink.

Table 5

SERUM URIC LEVELS BEFORE AND 30 TO 180
MIN FOLLOWING A SUCROSE LOAD (1 g/kg
BODY WEIGHT) IN YOUNG WOMEN AFTER
CONSUMING A DIET CONTAINING 43% OF
CALORIES AS EITHER SUCROSE OR STARCH
FOR 3 WEEKS EACH

Time (min)	Serum uric acid levels (mg %)			
	Women using oral contraceptives		Women not using oral contraceptives	
	Sucrose	Starch	Sucrose	Starch
0 (fasting)	4.2 ± 0.3	3.5 ± 0.3	3.9 ± 0.5	3.7 ± 0.2
30	4.3 ± 0.2	3.8 ± 0.3	3.9 ± 0.2	3.6 ± 0.4
60	4.1 ± 0.3	3.9 ± 0.2	4.1 ± 0.4	3.7 ± 0.3
120	4.1 ± 0.3	4.1 ± 0.2	4.2 ± 0.4	3.3 ± 0.3
180	3.6 ± 0.3	3.7 ± 0.2	4.4 ± 0.6	3.2 ± 0.2

Note: Each value represents the mean ± SEM from six women. ANOVA: Diet — $p < 0.005$; time — not significant; oral contraceptive — not significant.

Adapted from Kelsay, J. L., Behall, K. M., Moser, P. B., and Prather, E. S., *Am. J. Clin. Nutr.*, 30, 2016, 1977.

3. Extended Feeding

The feeding of fructose or sucrose to humans, either along or in comparison to other sources of dietary carbohydrate for extended periods of time, has not consistently increased the levels of blood uric acid. These apparently inconsistent findings could usually be explained by differences in the amount of fructose fed (dose dependency) or by the metabolic characteristics of the subjects studied.

The addition of 100 g of fructose to the usual diet of normal subjects for 5 days was reported to produce no significant increase in either plasma levels or the 24-hr urinary excretion of uric acid.[44] This amount of fructose represented a two- to fourfold increase in the daily intake. In contrast, when two patients with gout consumed a low-fructose diet for 1 year, they reported no acute attacks of gout and their serum uric acid response to a fructose load of 1 g/kg body weight was reduced to that observed for normal subjects.[57] When these subjects were given a diet rich in fructose for 3 weeks, the serum uric acid levels in response to a fructose load were again elevated.

A number of studies have compared the effects of diets containing either fructose or sucrose with those containing glucose-based carbohydrates on uric acid levels in both normal subjects and those with specific metabolic characteristics. The addition of 250 to 290 g of fructose as compared to an equal amount of glucose to the diet of three healthy men for 7 days in a crossover pattern produced significant increases in serum uric acid levels and 24-hr urinary excretion of uric acid in two of the three subjects.[60] These levels of fructose were extremely high and represented about 50% of the total calories. Kelsay et al.[61] have reported on the effect of feeding diets containing 43% of the calories from sucrose as compared to starch for 4 weeks each in a crossover design on serum uric acid levels in 12 healthy young women, on whom 6 were using oral contraceptives and 6 were not. The diets contained an average of 51% of calories from carbohydrate, 36% from fat, and 13% from protein. The uric acid levels before and 30 to 180 min following an oral load of 1 g sucrose per kilogram body weight after the 3rd week of each dietary period are presented in Table 5. Fasting

Table 6
SERUM URIC ACID LEVELS BEFORE AND 30 TO 180 MIN FOLLOWING A SUCROSE LOAD (2 g/kg BODY WEIGHT) IN MIDDLE-AGED MEN AND WOMEN AFTER CONSUMING A DIET CONTAINING 30% OF CALORIES AS EITHER SUCROSE OR WHEAT STARCH FOR 6 WEEKS EACH

| | Serum uric acid levels (mg %) | | | |
| | Sucrose | | Starch | |
Time (min)	Men	Women	Men	Women
0 (fasting)	6.7 ± 0.4	4.3 ± 0.3	6.2 ± 0.4	4.3 ± 0.2
30	7.1 ± 0.4	5.2 ± 0.3	6.5 ± 0.5	4.8 ± 0.3
60	7.0 ± 0.4	5.1 ± 0.3	6.6 ± 0.4	5.0 ± 0.3
120	7.1 ± 0.4	5.0 ± 0.3	6.4 ± 0.4	4.6 ± 0.3
180	6.5 ± 0.4	4.6 ± 0.3	6.0 ± 0.4	4.4 ± 0.3

Note: Each value represents the mean ± SEM from ten men and nine women. ANOVA: Diet — $p < 0.01$; sex — $p < 0.01$; diet × time — not significant; diet × sex — not significant.

Adapted from Solyst, J. T., Michaelis, O. E., IV, Reiser, S., Ellwood, K. C., and Prather, E. S., *Nutr. Metab.*, 24, 182, 1980.

serum uric acid levels were significantly higher at all times after the women consumed the sucrose as compared to the starch diet for 3 weeks. However, there was no additional increase in serum uric acid levels following the sucrose load in the subjects when they were fed sucrose as compared to starch. Serum uric acid levels were not affected by the use of oral contraceptives.

A long-term study (2 years) in which healthy subjects consumed sucrose, fructose, or xylitol on a self-selected diet reported no changes in serum uric acid in the xylitol and fructose groups.[62] There was a 12% increase of serum uric acid in the sucrose group, but this was not significant. Based on consumption records kept by the subjects, their average daily intake was 71, 68, and 48 g of sucrose, fructose, and xylitol, respectively. However, a decline in consumption occurred toward the end of the trial, and during the last 8 months the average daily intake was 58, 55, and 42 g for the respective sugars. During this time span, the average per capita intake of sucrose in the U.S. was 120 g/day, almost double that consumed in this study.[63] Consequently, the intake of these sugars actually may have been lower while the subjects were participating in the study than when they were consuming their previous diets, so that the effect may have been an overall decrease in consumption of sugar. Moreover, the intake of sucrose, fructose, and xylitol would be expected to increase blood uric acid levels.[50]

The effects on serum uric acid levels were determined after ten men and nine women 35 to 55 years of age and free of overt disease consumed diets containing 30% of calories from either sucrose or wheat starch for 6 weeks each in a crossover design.[64] The diets provided 43% of calories from carbohydrate, 42% from fat, and 15% from protein. Table 6 presents the effects of diet and sex on fasting serum uric acid levels and on levels 30 to 180 min after a sucrose load of 2 g/kg body weight. Serum uric acid levels were significantly greater ($p < 0.01$) at all times after the subjects consumed sucrose than starch. Uric acid values were also significantly greater in men than women ($p < 0.01$). However, serum uric acid

Table 7
SERUM URIC ACID
LEVELS BEFORE AND 2,
4, AND 7 DAYS AFTER
FOUR OBESE SUBJECTS
WERE FED EITHER 2000
kcal/DAY SUCROSE OR
3000 kcal/DAY GLUCOSE
FOR 7 DAYS

Day	Serum uric acid levels (mg %)	
	Sucrose	Glucose
0	6.3 ± 1.7	6.3 ± 1.3
2	7.0 ± 1.8	6.5 ± 1.0
4	7.9 ± 2.0[a]	6.6 ± 0.5
7	8.2 ± 2.6[a]	6.6 ± 0.9

Note: Each value is the mean $\pm$ S.D. from four subjects.

[a] Significantly different ($p < 0.05$) from values on day 0 as determined by analysis of variance.

Adapted from Fox, I. H., John, D., DeBruyne, S., Dwosh, I., and Marliss, E. B., *Metabolism,* 34, 741, 1985.

levels did not differ significantly in response to the sucrose load in subjects as a function of diet (diet $\times$ time interaction) or sex (sex $\times$ time interaction). These results are consistent with those observed after higher levels of sucrose (43% of calories) were fed for shorter time periods.[41]

In contrast to the results reported using controlled conditions and healthy subjects,[61,64] no significant changes in fasting serum uric acid were observed after eight young men consumed a low-protein, low-fat diet containing 35% of calories from sucrose as compared to dextromaltose in a crossover pattern for 3 weeks each.[65] The use of dextromaltose instead of starch as the comparatory carbohydrate would not be expected to increase uric acid levels.[44,58,59]

Fox et al.[19] compared the effects of feeding four obese subjects either 2000 kcal/day sucrose or 3000 kcal/day glucose for 7 days. The resultant serum uric acid levels are shown in Table 7. The subjects receiving the sucrose diet showed a significant increase in serum uric acid above pretest levels after the 4th day. The subjects receiving the glucose diet showed no significant increase in fasting serum uric acid above pretest levels during the 7-day feeding period. In another part of this study, two normal subjects and three subjects with gout were given a weight-maintaining diet containing 100% sucrose for an average of 9 days and serum uric acid was compared to base line levels. Uric acid in the normal subjects increased from a base line average of 4.7 to 8.5 mg % after consumption of the sucrose diet. The corresponding serum uric acid levels in the gouty subjects were 8.5 and 10.6 mg %.

The effect of sucrose intake on blood uric acid levels in 12 men and 12 women, classified as carbohydrate-sensitive on the basis of an exaggerated insulin response to a sucrose load,[66] has been recently determined.[67] The subjects consumed diets containing 5, 18, or 33% of

Table 8
**MEAN WEEKLY FASTING SERUM URIC
ACID AND MEAN URIC ACID
RESPONSE 30 TO 180 MIN FOLLOWING
A SUCROSE LOAD OF 2 g/kg BODY
WEIGHT OF 12 CARBOHYDRATE-
SENSITIVE MEN AND WOMEN
CONSUMING DIETS CONTAINING 5, 18,
OR 33% OF CALORIES AS SUCROSE**

| | Mean (mg %) | | | |
| | Fasting serum uric acid | | Uric acid response | |
Dietary sucrose (% of calories)	Men	Women	Men	Women
5	6.01[a]	4.40[c]	5.78[a]	4.04[c]
18	6.80[b]	4.74[d]	6.55[b]	4.37[c,d]
33	7.01[b]	5.00[d]	5.72[b]	4.56[d]

Note: Mean values with different superscript letters are significantly different from each other ($p < 0.05$) as determined by Duncan's multiple range test.

Adapted from Israel, K. D., Michaelis, O. E., IV, Reiser, S., and Keeney, M., *Ann. Nutr. Metab.*, 27, 425, 1983.

calories as sucrose at the expense of wheat starch in a crossover pattern for 6 weeks each. The diets contained 44% of calories from carbohydrate, 42% from fat, and 14% from protein. Table 8 presents the mean weekly fasting serum uric acid and mean uric acid response 30 to 180 min following a sucrose load of 2 g/kg body weight as a function of the sucrose content of the diets and the sex of the subject. Mean fasting uric acid levels were significantly greater in both men and women while they consumed either the 18 or 33% than the 5% sucrose diet. Fasting serum uric acid and mean uric acid responses were higher in men than in women at all levels of sucrose. Uric acid response to the sucrose load was greater in men after they consumed either the 18 or 33% than the 5% sucrose diet, and in women after they consumed the 33% than the 5% sucrose diet. In a related study, the effect of various levels of dietary fructose on blood lipid levels of 12 hyperinsulinemic men and 12 normoinsulinemic controls was determined.[68] Diets providing 0, 7.5, or 15% of the calories from fructose at the expense of wheat starch were fed for 5 weeks each in a crossover pattern. Plasma uric acid levels of the hyperinsulinemic men were about 20% higher than those of the controls ($p = 0.05$) 3 hr after a sucrose load of 2 g/kg body weight. Uric acid levels were an average of 22% greater ($p < 0.05$) in hyperinsulinemic men 5 weeks after they consumed the 7.5 or 15% than the 0% fructose diet.

In contrast to the findings with hyperinsulinemic subjects, it was reported that the feeding of liquid formula diets containing either 9 or 17% of the total calories from fructose at the expense of dextromaltose for about 2 weeks to six hypertriglyceridemic men resulted in no change in fasting serum uric acid levels.[69] Because of the positive relationship between blood triglyceride and blood insulin levels,[66] it would have been expected that these subjects would have reacted to fructose in a way similar to the hyperinsulinemic subjects.[67,68] Differences in the experimental design used in these studies may explain the apparently conflicting results. For example, in the study with the hypertriglyceridemic subjects the liquid formula diet was given in equal portions five times throughout the day.[69] In contrast, in the studies with the hyperinsulinemic subjects the diets were fed either two[67] or three[68] times daily, with 55 to 75% of the calories consumed during dinner. It has been shown that consumption of calories in a gorging as compared to a nibbling pattern can increase levels of blood insulin[70] and triglycerides.[71]

4. Conclusions

These results indicate that under specified conditions (such as i.v. infusion or the oral consumption of large amounts of sucrose or fructose) by which blood fructose levels could be raised acutely, blood uric acid levels are increased. Since high levels of uric acid are associated with a number of disease states, these high levels of blood fructose are metabolically undesirable. The nutritional significance of these types of studies is questionable, since fructose or sucrose is usually consumed in smaller quantities as part of a mixed meal. Under these conditions, the levels of fructose reaching the liver would be much smaller than those achieved by infusion. However, consumption of fructose or sucrose at levels approximating those presently consumed in this country, particularly in a gorging pattern, may provoke increases in blood uric acid in certain population groups such as those with gout or metabolic defects leading to hyperinsulinemia. These effects may be not only acute but also adaptive, as a result of the increase in the activities of intestinal sucrase digesting sucrose and hepatic enzymes metabolizing fructose, when the diet contains fructose and/or sucrose.

C. Mechanisms for Fructose-Induced Hyperuricemia

1. Introduction

Two general mechanisms have been proposed to explain the ability of fructose to increase uric acid levels. The one most generally accepted proposes that fructose catabolism in the liver produces changes in the metabolic environment conducive to the breakdown of adenine nucleotides to uric acid. The second envisions a stimulatory influence of fructose on *de novo* purine synthesis in the liver. The experimental evidence in support of both mechanisms will be presented in this section.

2. Adenine Nucleotide Breakdown

Many studies using experimental animals have confirmed that the administration of fructose produces a rapid decrease in the level of hepatic ATP.[72-80] This finding was first reported by Mäenpää et al.[72] Fructose was injected intravenously into anesthetized rats and a precipitous depletion of ATP was observed within 1 min. The decrease in ATP was not followed by an equivalent increase in the sum of ADP + AMP. When glucose, galactose, ribose, or sorbitol was injected instead of fructose, only sorbitol decreased ATP and total adenine nucleotides; however, its effect was quantitatively much smaller than that of fructose. The level of plasma uric acid and allantoin increased about fourfold within 15 min of the injection of fructose, indicating that breakdown of the adenine nucleotides was the source of the uric acid and allantoin.

Studies investigating the mechanism of the fructose-induced degradation of adenine nucleotides have been performed using freeze-clamped livers from rats previously perfused with fructose.[75,76] The main effect observed after fructose perfusion was the accumulation of fructose-1-phosphate as the result of the phosphorylation of fructose by fructokinase. Other metabolic effects observed were a 35% decrease in the concentration of total adenine nucleotides, a 23% loss of ATP content, a decrease in inorganic phosphate, and a sevenfold rise in the concentration of inosine monophosphate. The decrease in the levels of ATP and inorganic phosphate was attributed to the sequestration of these compounds in the formation of fructose-1-phosphate from fructose. The high levels of fructose-1-phosphate were also attributed to an inhibition of fructose-1-phosphate aldolase by inosine monophosphate; however, the importance of this inhibition has subsequently been questioned.[77]

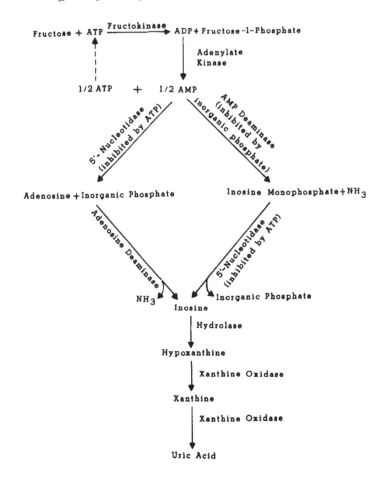

FIGURE 2.   Metabolic events leading to the fructose-induced increase in uric acid production in mammalian liver.

The importance of the levels of inorganic phosphate on hepatic adenine nucleotide metabolism has been shown in a study in which rats were intraperitoneally injected with either fructose alone or fructose in combination with inorganic phosphate.[78] With fructose loading alone, levels of ATP, total adenine nucleotides, and inorganic phosphate fell. In contrast, prior phosphate loading attenuated the fall of ATP and total adenine nucleotides characteristic of fructose loading. It was concluded that not only was inorganic phosphate a factor in the fall of ATP and total adenine nucleotides, but also the severity of the depletion of inorganic phosphate determined the severity of reductions in ATP and total adenine nucleotides.

The levels of hepatic ATP and inorganic phosphate appear to control the rate of conversion of AMP to inosine, since inorganic phosphate inhibits AMP deaminase[81] and ATP inhibits 5′-nucleotidase.[82] The increase in uric acid observed after fructose administration can therefore be explained on the basis of a fructose-induced reduction of both hepatic ATP and inorganic phosphate, leading to a release of the physiological inhibition of both the deaminase and the nucleotidase. The metabolic events leading to the fructose-induced increase in uric acid production in mammalian liver are summarized in Figure 2.

The infusion of fructose has also been shown to decrease the adenine nucleotide content of human liver.[83] Table 9 shows the ATP, ADP, AMP, and total adenine nucleotide content of liver biopsy specimens before and after the perfusion of 500 mℓ of either 10% fructose or glucose for 30 min. The infusion of fructose produced significant decreases ($p < 0.05$) of 40 to 50% from initial levels of liver ATP, ADP, and total nucleotides; the AMP content

Table 9

**LIVER CONTENT OF ATP, ADP, AMP, AND TOTAL ADENINE
NUCLEOTIDES BEFORE AND AFTER HUMANS WERE INFUSED
WITH 500 mℓ OF EITHER 10% FRUCTOSE OR GLUCOSE FOR 30 MIN**

Experimental conditions	Adenine nucleotide content (μmol/g liver)			
	ATP	ADP	AMP	Total
Before fructose ($n = 6$)	1.91 ± 0.56	0.98 ± 0.25	0.96 ± 0.36	3.90 ± 0.36
After fructose ($n = 6$)	1.04 ± 0.55[a]	0.69 ± 0.15[a]	0.89 ± 0.16	2.62 ± 0.53[a]
Before glucose ($n = 2$)	2.59 ± 2.11	0.99 ± 0.92	0.81 ± 0.61	4.39 ± 3.64
After glucose ($n = 2$)	2.24 ± 1.38	1.06 ± 1.19	0.79 ± 0.95	4.09 ± 3.52

Note: Each value is the mean ± S.D. from the number of subjects shown in parentheses.

[a] Significantly lower ($p < 0.05$) than corresponding level before infusion.

Adapted from Bode, C., Schumacher, H., Goebell, H., Zelder, O., and Pelzel, H., *Horm. Metab. Res.,* 3, 289, 1971.

was not affected by fructose infusion. No significant changes in the hepatic content of any of the adenine nucleotides were observed after the infusion of glucose. In contrast to these results, the ATP content of erythrocytes was not decreased after gouty patients were infused with 0.5 g fructose per kilogram body weight.[52]

3. De Novo Purine Biosynthesis

A number of studies have suggested that fructose administration may raise uric acid levels through an increase in purine biosynthesis. Using mice, it was reported that the i.v. injection of 2.5 mg/g body weight of fructose as compared to glucose led to a threefold increase in the incorporation of ^{14}C glycine into hepatic purine nucleotides.[84] In the mice treated with fructose, but not glucose, a reduction in the hepatic content of nucleotides that inhibit amidophosphoribosyltransferase, the enzyme catalyzing the first reaction of purine biosynthesis, was observed. In addition, the metabolic availability of phosphoribosyl pyrophosphate, a key substrate in the biosynthesis of purine nucleotides, was increased 2.3-fold in the liver after fructose, but not glucose, administration. Indirect evidence for a fructose-activated synthesis of uric acid was reported in a study using rats. Using oxonate to inhibit the conversion of uric acid to allantoin, it was reported that fructose infusion produced an increase in serum uric acid while not significantly decreasing the hepatic concentration of ATP.[80]

Evidence for the acceleration of *de novo* purine synthesis by fructose also has been obtained in human studies. The effect of the i.v. infusion of 125 to 200 g of fructose over a period of 3 to 4 hr on the incorporation of 1-^{14}C-glycine into urinary urate was determined in four subjects with gout and one with osteoarthritis.[53] The incorporation of radioactive glycine was increased from 21 to 430% by fructose infusion. It was not clear whether the increased rate of purine synthesis was in response to the fructose-induced depletion of purine nucleotides or to a direct stimulation by a fructose metabolite. A fructose-induced increase in the important regulator of nucleotide synthesis, phosphoribosyl pyrophosphate, appeared unlikely since the level of this substrate in erythrocytes was found to be decreased by fructose in this study as well as in a similar study by Fox and Kelley.[52] The oral intake of large quantities of fructose also appears to increase uric acid synthesis. The feeding of diets

Table 10
LACTATE FORMATION
BY PARENCHYMAL
CELLS ISOLATED FROM
THE LIVERS OF FASTED
RATS 1 HR AFTER
INCUBATION WITH
FRUCTOSE AS
COMPARED TO GLUCOSE

Substrate	Lactate formed (μmol/g/hr)
None	-3.7 ± 1.3
10 mM fructose	116.3 ± 11.0
10 mM glucose	6.0 ± 9.9
20 mM fructose	99.3 ± 6.4
20 mM glucose	13.7 ± 6.9

Note: Each value is the mean $\pm$ SEM from three cell samples.

Adapted from Seglin, P. O., *Biochim. Biophys. Acta*, 338, 317, 1974.

containing 250 to 290 g of fructose as compared to glucose for at least 5 days produced a nearly twofold increase in the incorporation of radioactive glycine into urinary urate in three men with stable chronic neurological disease.[60]

III. LACTIC ACID

Lactic acid is the end product of anaerobic glycolysis that occurs in most mammalian tissues. It is also a normal component of human blood. The confirmed normal range of blood lactic acid is 9.0 to 10.1 mg %; however, considerable variations about this confirmed range have been reported.[1] High levels of blood lactic acid in humans produce the extremely toxic condition lactic acidosis.

A. Studies with Rats

Using various preparations of rat liver, it has been demonstrated repeatedly that lactate formation occurs much more readily from fructose than from glucose. The incubation of liver slices from normal rats with 30 mM fructose as compared to glucose resulted in the production of four times as much lactate after 90 min.[85] In the same study it was found that liver slices from alloxan-diabetic rats produced two to four times more lactate from fructose than from glucose. Similarly the incubation of 11 mM fructose with liver slices from young rats resulted in the formation of about three times more lactate than 1 hr as compared to glucose.[86] Perfusion of liver from fasted young rats with 20 mM fructose resulted in the production of 46 μmol of lactate per gram liver per hour.[87] In contrast, no net formation of lactate was observed after the perfusion of 12 mM glucose under the same experimental conditions. The amount of lactate formed from equal concentrations of fructose and glucose by parenchymal cells isolated from the livers of fasted rats is shown in Table 10.[88] The amount of lactate formed from fructose was 7- to 19-fold greater than that formed from glucose. The more rapid formation of lactate from fructose than from glucose can be generally explained by (1) greater capacity of fructokinase to phosphorylate fructose than of hexokinase and glucokinase to phosphorylate glucose;[89] (2) the by-passing of the regulatory

Table 11
PLASMA LACTATE LEVELS
BEFORE AND 15 TO 90 MIN AFTER
THE INGESTION OF 1 g/kg BODY
WEIGHT OF GLUCOSE, SUCROSE,
OR FRUCTOSE BY HEALTHY MEN

Time (min)	Plasma lactate levels (mg %)		
	Glucose	Sucrose	Fructose
0	16.2 ± 5.4	9.0 ± 1.3	10.5 ± 1.3
15	11.3 ± 3.2	11.1 ± 1.9	11.7 ± 1.0
30	10.9 ± 1.0	17.5 ± 2.0	19.1 ± 1.1
60	12.5 ± 0.8	20.0 ± 1.8	18.1 ± 1.2
90	11.7 ± 0.8	15.7 ± 1.7	14.1 ± 0.6

Note: Each value represents the mean ± SEM from nine men.

Adapted from Macdonald, I., Keyser, A., and Pacy, D., *Am. J. Clin. Nutr.*, 31, 1305, 1978.

step in glycolysis, the reaction catalyzed by phosphofructokinase; and (3) the stimulation of pyruvate kinase by fructose-1-phosphate.[90] Alternatively, fructose may activate glycolysis from preformed glycogen in tissues such as the liver and muscle.

B. Studies with Humans
1. Infusion Studies

The infusion of 0.5 g fructose per kilogram body weight into both normal children and those with hereditary fructose intolerance provoked a large and rapid increase in blood lactic acid levels.[41] The fructose-induced increase in lactic acid was explained by a block in gluconeogenesis caused by the inhibition of glucose-phosphate isomerase by fructose-1-phosphate. Serum lactate levels 90 min after ten healthy men were infused with a 10% fructose solution calculated to provide 0.5 g/kg body weight rose significantly ($p < 0.02$) from a preinfusion average of 16.3 mg % to a peak concentration of 62.2 mg %.[45] The peak level of serum fructose occurred at this time. It was concluded that fructose-induced lactate acidosis may exacerbate already established acidosis in patients with diabetes, uremia, anoxia, or postoperative stress. Significant increases in blood lactic acid levels were observed in 17 normal subjects and 24 diabetics following the infusion of a 40% fructose solution calculated to provide 0.5 g/kg ideal body weight over a 4-min period.[54] Lactic acid levels in the diabetics were about 50% higher than those in the normal subjects. The infusion of 0.6 g fructose per kilogram body weight per hour for 3 hr into 30 patients free of diabetes or liver disease resulted in a steady-state increase of 2- to 3.5-fold in blood lactate concentrations $1\frac{1}{2}$ to 3 hr after the start of the infusion.[91] Plasma lactate levels were increased about 2.3-fold 15 min after eight normal subjects were infused with 0.5 g fructose per kilogram body weight;[44] 30 min after a similar infusion into four men with gout, blood lactate concentration was reported to increase an average of 3.4-fold over control levels.[52]

2. Feeding Studies

A significant increase in blood lactic acid levels in response to an oral load of fructose, but not glucose, has been reported in two studies. The mean blood lactate levels after the intake of 50 g of fructose were approximately three times greater than after the intake of an equal amount of glucose.[92] Table 11 presents the results of a study in which nine healthy

Table 12

**WHOLE-BLOOD LACTATE LEVELS BEFORE AND
30 TO 180 MIN FOLLOWING A SUCROSE LOAD (1
g/kg BODY WEIGHT) IN YOUNG WOMEN AFTER
CONSUMING A DIET CONTAINING 43% OF
CALORIES AS EITHER SUCROSE OR STARCH
FOR 3 WEEKS EACH**

	Whole-blood lactate levels (mg %)			
Time (min)	Women using oral contraceptives		Women not using oral contraceptives	
	Sucrose	Starch	Sucrose	Starch
0 (fasting)	8.2 ± 1.3	8.1 ± 1.1	7.8 ± 0.6	7.6 ± 1.2
30	19.5 ± 2.2	15.5 ± 2.8	11.3 ± 2.1	11.7 ± 1.7
60	22.5 ± 2.1	19.2 ± 1.6	11.3 ± 1.7	12.8 ± 2.3
120	10.7 ± 1.4	13.2 ± 1.3	8.0 ± 0.8	8.7 ± 0.6
180	8.1 ± 0.7	9.3 ± 1.1	7.2 ± 0.6	7.8 ± 0.4

Note: Each value represents the mean ± SEM from six women. ANOVA: Diet — not significant; time — $p < 0.005$; oral contraceptive — $p < 0.005$.

Adapted from Kelsay, J. L., Behall, K. M., Moser, P. B., and Prather, E. S., *Am. J. Clin. Nutr.*, 30, 2016, 1977.

men were given 1 g/kg body weight of glucose, sucrose, or fructose.[58] A significant fall ($p < 0.025$) in mean serum lactate levels was observed after the intake of glucose. In contrast, a significant rise in mean serum lactate was observed after the intake of sucrose or fructose. The increase after sucrose intake must therefore be due to the fructose moiety of sucrose.

The effect of the extended feeding of sucrose as compared to glucose-based carbohydrates on blood lactic acid levels has been reported in three studies. Table 12 presents whole-blood lactate levels before and 30 to 180 min following a sucrose load (1 g/kg body weight) in 12 young women after they consumed a diet containing 43% of calories as either sucrose or starch for 3 weeks each;[61] 6 of the women were using oral contraceptives and 6 were not. The nature of the dietary carbohydrate did not significantly affect the magnitude of the increase in blood lactate provoked by the sucrose load. The women using oral contraceptives responded to the sucrose load with significantly higher lactate responses than did the women not using oral contraceptives ($p < 0.005$). In a somewhat similar study, serum lactic acid levels were determined before and 30 to 180 min after a sucrose load (2 g/kg body weight) in 19 middle-aged subjects after they consumed a diet containing 30% of calories as either sucrose or wheat starch for 6 weeks each.[64] The results are shown in Table 13. The sucrose load again produced a significant increase in serum lactic acid levels, and again this increase was not significantly influenced by the nature of the dietary carbohydrate fed prior to the test load.

In a study in which four obese, nongouty subjects were fed a diet containing either 2000 kcal sucrose or 3000 kcal glucose daily for 7 days, the increase in blood lactate above pretest levels was about the same (1.8- and 2.0-fold for sucrose and glucose, respectively).[19] This observation appears to be contradictory to previous findings reporting that fructose or sucrose produces significantly greater increases in blood lactate levels than does glucose. The much higher caloric intake of the subjects while consuming the glucose as compared to the sucrose diets may partially explain these findings.

Table 13
SERUM LACTIC ACID LEVELS BEFORE AND 30 TO 180 MIN FOLLOWING A SUCROSE LOAD (2 g/kg BODY WEIGHT) IN MIDDLE-AGED SUBJECTS (TEN MEN AND NINE WOMEN) AFTER CONSUMING A DIET CONTAINING 30% OF CALORIES AS EITHER SUCROSE OR WHEAT STARCH FOR 6 WEEKS EACH

Time (min)	Serum lactic acid levels (mg %)	
	Sucrose	Starch
0 (fasting)	6.9 ± 0.8	6.8 ± 1.0
30	11.6 ± 1.3	9.8 ± 1.0
60	13.4 ± 0.8	14.0 ± 1.3
120	9.5 ± 0.8	11.5 ± 1.3
180	6.6 ± 0.6	6.7 ± 0.7

Note: Each value represents the mean ± SEM from 23 subjects. ANOVA: Sex — not significant; diet — not significant; time — $p < 0.01$; interactions — not significant.

Adapted from Solyst, J. T., Michaelis, O. E., IV, Reiser, S., Ellwood, K. C., and Prather, E. S., *Nutr. Metab.*, 24, 182, 1980.

REFERENCES

1. **Lippert, H. and Lehmann, H. P.**, *SI Units in Medicine*, Urban & Schwarzenberg, Baltimore, 1978, 1.
2. **Watts, R. W. E., Kelley, W. N., Rapado, A., Scott, J. T., Seegmiller, J. E., deVries, A., Wyngaarden, J. B., and Zollner, N.**, Panel discussion: hyperuricemia as a risk factor, *Adv. Exp. Med. Biol.*, 76B, 342, 1977.
3. **Gertler, M. M., Garn, S. M., and Levine, S. A.**, Serum uric acid in relation to age and physique in health and in coronary heart disease, *Ann. Intern. Med.*, 34, 1421, 1951.
4. **Kohn, P. N. and Prozan, G. B.**, Hyperuricemia — relationship to hypercholesterolemia and acute myocardial infarction, *JAMA*, 170, 1909, 1959.
5. **Jacobs, D.**, Hyperuricemia and myocardial infarction, *S. Afr. Med. J.*, 46, 367, 1972.
6. **Srivastava, B. N., Om, H., Deshmankar, B. S., and Gupta, R. K.**, Serum uric acid in coronary heart disease, *Ind. Heart J.*, 26, 24, 1974.
7. **Yano, K., Rhoads, G. G., and Kagan, A.**, Epidemiology of serum uric acid among 8000 Japanese-American men in Hawaii, *J. Chron. Dis.*, 30, 171, 1977.
8. **Jacobs, D. R. and Bernstein, W. H.**, Serum uric acid: its association with other risk factors and with mortality in coronary heart disease, *J. Chron. Dis.*, 29, 557, 1976.
9. **Takkunen, H. and Reunanen, A.**, Hyperuricemia and other cardiovascular risk factors, *Adv. Exp. Med. Biol.*, 76B, 238, 1977.
10. **Schoenfeld, M. R. and Goldberger, E.**, Serum cholesterol-uric acid correlations, *Metabolism*, 12, 714, 1963.
11. **Cannon, P. J., Stason, W. B., Demartini, F. E., Sommers, S. C., and Laragh, J. H.**, Hyperuricaemia in primary and renal hypertension, *N. Engl. J. Med.*, 275, 457, 1966.

12. **Muller, G., Seige, K., and Stammberger, K.,** Serum uric acid in diabetes, *Dtsch. Gesundheitswes.*, 33, 2176, 1978.
13. **Ballantyne, D., Strevens, E. A., and Lawrie, T. D. V.,** Relationship of plasma uric acid to plasma lipids and lipoproteins in subjects with peripheral vascular disease, *Clin. Chim. Acta*, 70, 323, 1976.
14. **Jensen, J., Blackenhorn, D. H., and Kornerup, V.,** Blood uric acid levels in familial hypercholesterolaemia, *Lancet*, 2, 298, 1966.
15. **Bansal, B. C., Gupta, R. R., Bansal, M. R., and Prakash, C.,** Serum lipids and serum uric acid relationship in ischemic thrombotic cerebrovascular disease, *Stroke*, 6, 304, 1975.
16. **Barlow, K. A.,** Hyperlipidemia in primary gout, *Metabolism*, 17, 289, 1968.
17. **Wyngaarden, J. B. and Kelley, W. N.,** *Gout and Hyperuricemia*, Grune & Stratton, Orlando, Fla., 1976, 21.
18. **Berkowitz, D.,** Blood lipid and uric acid interrelationships, *JAMA*, 190, 120, 1964.
19. **Fox, I. H., John, D., DeBruyne, S., Dwosh, I., and Marliss, E. B.,** Hyperuricemia and hypertriglyceridemia: metabolic basis for the association, *Metabolism*, 34, 741, 1985.
20. **Bluestone, R., Lewis, B., and Mervart, I.,** Hyperlipoproteinaemia in gout, *Ann. Rheum. Dis.*, 30, 134, 1971.
21. **Alepa, F. P., Howell, R. R., Klinenberg, J. R., and Seegmiller, J. E.,** Relationships between glycogen storage disease and tophaceous gout, *Am. J. Med.*, 42, 58, 1967.
22. **Kelley, W. N., Rosenbloom, F. M., Seegmiller, J. E., and Howell, R. R.,** Excessive production of uric acid in type I glycogen storage disease, *J. Pediatr.*, 72, 488, 1968.
23. **Greene, H. L., Wilson, F. A., Hefferan, P., Terry, A. B., Moran, J. R., Slonim, A. E., Claus, T. H., and Burr, I. M.,** ATP depletion, a possible role in the pathogenesis of hyperuricemia in glycogen storage disease type I, *J. Clin. Invest.*, 62, 321, 1978.
24. **Pagliara, A. S., Karl, I. E., Keeting, J. P., Brown, B. I., and Kipnis, D. M.,** Hepatic fructose-1,6-diphosphatase deficiency: a cause for lactic acidosis and hypoglycemia in infancy, *J. Clin. Invest.*, 51, 2115, 1972.
25. **Ramsay, L. E.,** Hyperuricaemia in hypertension: role of alcohol, *Br. Med. J.*, 1, 653, 1979.
26. **Fessel, W. J., Siegelaub, A. B., and Johnson, E. S.,** Correlates and consequences of asymptomatic hyperuricemia, *Arch. Intern. Med.*, 132, 44, 1973.
27. **Rapado, A. and Castrillo, J. M.,** Gout disease. Its natural history based on 1000 observations, *Int. J. Vitam. Nutr. Res. Suppl.*, 15, 223, 1976.
28. **Anisimov, V. E., Zheltukhina, V. V., and Semavin, I. E.,** Content of uric acid in coronary arteriosclerosis and hypertensive disease, *Klin. Med.*, 54, 29, 1976.
29. **Breckenridge, A.,** Hypertension and hyperuricaemia, *Lancet*, 1, 15, 1966.
30. **Fiaschi, E., Baggio, B., Favaro, S., Antonello, A., Camerin, E., Todesco, S., and Borsatti, A.,** Fructose-induced hyperuricemia in essential hypertension, *Metabolism*, 26, 1219, 1977.
31. **Pearce, J. and Aziz, H.,** Uric acid and plasma lipids in cerebrovascular disease. I. Prevalence of hyperuricaemia, *Br. Med. J.*, 4, 78, 1969.
32. The Coronary Drug Project Research Group, Serum uric acid: its association with other risk factors and with mortality in coronary heart disease, *J. Chron. Dis.*, 29, 557, 1976.
33. **Paulus, H. E., Coutts, A., Calabro, J. J., and Klineberg, J. R.,** Clinical significance of hyperuricemia in routinely screened hospitalized men, *JAMA*, 211, 277, 1970.
34. **Al-Hujaj, M., Schönthal, H., and Elbrechter, J.,** Differential diagnosis of gout, *Lancet*, 2, 606, 1972.
35. **Hall, A. P., Barry, P. E., and Dawber, T. R.,** Epidemiology of gout and hyperuricemia. A long-term population study, *Am. J. Med.*, 42, 27, 1967.
36. **Mikkelson, W. M.,** The possible association of hyperuricemia and/or gout with diabetes, *Arthritis Rheum.*, 8, 853, 1965.
37. **Reed, D., Labarthe, D., and Stallones, R.,** Epidemiologic studies of serum uric acid levels among Micronesians, *Arthritis Rheum.*, 15, 38, 1972.
38. **Skeith, M. D., Healey, L. A., and Cutler, R. E.,** Urate excretion during mannitol and glucose diuresis, *J. Lab. Clin. Med.*, 70, 213, 1967.
39. **Herman, J. B. and Keynan, A.,** Hyperglycemia and uric acid, *Isr. J. Med. Sci.*, 5, 1048, 1969.
40. **Herman, J. B., Medalie, J. H., and Goldbourt, N.,** Diabetes, prediabetes and uricaemia, *Diabetologia*, 21, 47, 1976.
41. **Perheentupa, J. and Raivio, K.,** Fructose-induced hyperuricaemia, *Lancet*, 2, 528, 1967.
42. **Al-Hujaj, M. and Schönthal, H.,** Fructose and hyperuricemia, *Metabolism*, 24, 899, 1975.
43. **Brodan, V., Brodanová, M., Kuhn, E., Filip, J., and Pechar, J.,** Ammonia and uric acid formation after rapid intravenous fructose administration to healthy human subjects and patients with compensated cirrhosis of the liver, *Nutr. Metab.*, 19, 233, 1975.
44. **Narins, R. G., Weisberg, J. S., and Meyers, A. R.,** Effects of carbohydrates on uric acid metabolism, *Metabolism*, 23, 455, 1974.

45. **Sahebjami, H. and Scalettar, R.,** Effects of fructose infusion on lactate and uric acid metabolism, *Lancet,* 1, 366, 1971.

46. **Curreri, P. W. and Pruitt, B. A., Jr.,** Absence of fructose-induced hyperuricaemia in men, *Lancet,* 1, 839, 1970.

47. **Heuckenkamp, P.-U. and Zollner, N.,** Fructose-induced hyperuricaemia, *Lancet,* 1, 808, 1971.

48. **Förster, H.,** Possible side effects of glucose, fructose, sorbitol and xylitol in man, *Int. J. Vitam. Nutr. Res. Suppl.,* 15, 116, 1976.

49. **Förster, H.,** Carbohydrates in parenteral nutrition, *Nutr. Metab.,* 20(Suppl. 1), 57, 1976.

50. **Förster, H. and Hoos, I.,** Carbohydrate-induced increase in uric acid synthesis. Studies in human volunteers and in laboratory rats, *Adv. Exp. Med. Biol.,* 76A, 519, 1977.

51. **Kogut, M. D., Roe, T. F., Won, W., and Donnell, G. N.,** Fructose-induced hyperuricemia: observations in normal children and in patients with hereditary fructose intolerance and galactosemia, *Pediatr. Res.,* 9, 744, 1975.

52. **Fox, I. H. and Kelley, W. N.,** Studies on the mechanism of fructose-induced hyperuricemia in man, *Metabolism,* 21, 713, 1972.

53. **Raivio, K. O., Becker, M. A., Meyer, L. J., Greene, M. L., Nuki, G., and Seegmiller, J. E.,** Stimulation of human purine synthesis de novo by fructose infusion, *Metabolism,* 24, 861, 1975.

54. **El-Ebrashy, W., Shaheen, M. H., Wasfi, A. A., and El-Danasoury, M.,** Side effects of IV fructose load in diabetics, *J. Egypt. Med. Assoc.,* 57, 406, 1974.

55. **Lewis, B.,** Primary hypertriglyceridaemic states, in *The Hyperlipidaemias: Clinical and Laboratory Practice,* Blackwell Scientific, Oxford, 1976, chap. 12.

56. **Marugo, M., Scopinaro, N., Minuto, F., and Barreca, T.,** Role of carbohydrates in the diet in patients with gout, *Boll. Soc. Ital. Biol. Sper.,* 46, 514, 1970.

57. **Stirpe, F., Corte, E. D., Bonetti, E., Abbondanza, A., Abbati, A., and de Stefano, F.,** Fructose-induced hyperuricaemia, *Lancet,* 2, 1310, 1970.

58. **Macdonald, I., Keyser, A., and Pacy, D.,** Some effects, in man, of varying the load of glucose, sucrose, fructose, or sorbitol on various metabolites in blood, *Am. J. Clin. Nutr.,* 31, 1305, 1978.

59. **Reiser, S., Scholfield, D., Trout, D., Wilson, A., and Aparicio, P.,** Effect of glucose and fructose on the absorption of leucine in humans, *Nutr. Rep. Int.,* 30, 151, 1984.

60. **Emmerson, B. T.,** Effect of oral fructose on urate production, *Ann. Rheum. Dis.,* 33, 276, 1974.

61. **Kelsay, J. L., Behall, K. M., Moser, P. B., and Prather, E. S.,** The effect of kind of carbohydrate in the diet and use of oral contraceptives on metabolism of young women. I. Blood and urinary lactate, uric acid, and phosphorus, *Am. J. Clin. Nutr.,* 30, 2016, 1977.

62. **Huttunen, J. K., Mäkinen, K. K., and Scheinin, A.,** Turku sugar studies. XI. Effects of sucrose, fructose and xylitol diets on glucose, lipid and urate metabolism, *Acta Odont. Scand.,* 33(Suppl. 70), 239, 1975.

63. **Friend, B. and Marston, R.,** Nutritional review, in National Food Situation, Economic Research Service, U.S. Department of Agriculture, Washington, D.C., 1974, 26.

64. **Solyst, J. T., Michaelis, O. E., IV, Reiser, S., Ellwood, K. C., and Prather, E. S.,** Effect of dietary sucrose in humans on blood uric acid, phosphorus, fructose, and lactic acid responses to a sucrose load, *Nutr. Metab.,* 24, 182, 1980.

65. **Richardson, D. P., Scrimshaw, N. S., and Young, V. R.,** The effect of dietary sucrose on protein utilization in healthy young men, *Am. J. Clin. Nutr.,* 33, 264, 1980.

66. **Reiser, S., Bickard, M. C., Hallfrisch, J., Michaelis, O. E., IV, and Prather, E. S.,** Blood lipids and their distribution in hyperinsulinemic subjects fed three different levels of sucrose, *J. Nutr.,* 111, 1045, 1981.

67. **Israel, K. D., Michaelis, O. E., IV, Reiser, S., and Keeney, M.,** Serum uric acid, inorganic phosphorus, and glutamic-oxalacetic transaminase and blood pressure in carbohydrate-sensitive adults consuming three different levels of sucrose, *Ann. Nutr. Metab.,* 27, 425, 1983.

68. **Hallfrisch, J., Ellwood, K., Michaelis, O. E., IV, Reiser, S., and Prather, E. S.,** Plasma fructose, uric acid, and inorganic phosphorus responses of hyperinsulinemic men fed fructose, *J. Am. Coll. Nutr.,* 5, 61, 1986.

69. **Turner, J. L., Bierman, E. L., Brunzell, J. D., and Chait, A.,** Effect of dietary fructose on triglyceride transport and glucoregulatory hormones in hypertriglyceridemic men, *Am. J. Clin. Nutr.,* 32, 1043, 1979.

70. **Pringle, D. J., Wadhwa, P. S., and Elson, C. E.,** Influence of frequency of eating low energy diets on insulin response in women during weight reduction, *Nutr. Rep. Int.,* 13, 339, 1976.

71. **Gwinup, G., Byron, R. C., Roush, W. H., Kruger, F. A., and Hamwi, G. J.,** Effect of nibbling versus gorging on serum lipids in man, *Am. J. Clin. Nutr.,* 13, 209, 1963.

72. **Mäenpää, P. H., Ravio, K. O., and Kekomaki, M. P.,** Liver adenine nucleotides: fructose-induced depletion and its effect on protein synthesis, *Science,* 161, 1253, 1968.

73. **Burch, H. B., Max, P., Jr., Chyu, K., and Lowry, O. H.,** Metabolic intermediates in liver of rats given large amounts of fructose or dihydroxyacetone, *Biochem. Biophys. Res. Commun.,* 34, 619, 1969.

74. **Burch, H. B., Lowry, O. H., Meinhardt, L., Max, P., Jr., and Chyu, K.-J.,** Effect of fructose, dihydroxyacetone, glycerol, and glucose on metabolites and related compounds in liver and kidney, *J. Biol. Chem.,* 245, 2092, 1970.

75. **Woods, H. F., Eggleston, L. V., and Krebs, H. A.,** The cause of hepatic accumulation of fructose 1-phosphate on fructose loading, *Biochem. J.,* 119, 501, 1970.

76. **Woods, H. F.,** Hepatic accumulation of metabolites after fructose loading, *Acta Med. Scand.,* Suppl. 542, 87, 1972.

77. **Van den Berghe, G., Hue, L., and Hers, H.-G.,** Effect of administration of fructose on the glycogenolytic action of glucagon, *Biochem. J.,* 134, 637, 1973.

78. **Morris, R. C., Jr., Nigon, K., and Reed, E. B.,** Evidence that the severity of depletion of inorganic phosphate determines the severity of the disturbance of adenine nucleotide metabolism in the liver and renal cortex of the fructose-loaded rat, *J. Clin. Invest.,* 61, 209, 1978.

79. **Itoh, R.,** Effect of oral administration of fructose on purine nucleotide metabolism in rats, *Comp. Biochem. Physiol.,* 26B, 817, 1983.

80. **Brosh, S., Boer, P., and Sperling, O.,** Effects of fructose on purine nucleotide metabolism in isolated rat hepatocytes, *Adv. Exp. Med. Biol.,* 165A, 481, 1984.

81. **Nikiforuk, G. and Colowick, S. P.,** The purification and properties of 5-adenylic acid deaminase from muscle, *J. Biol. Chem.,* 219, 119, 1956.

82. **Baer, H. P., Drummond, G. I., and Duncan, E. L.,** Formation and deamination of adenosine by cardiac muscle enzyme, *Mol. Pharmacol.,* 2, 67, 1966.

83. **Bode, C., Schumacher, H., Goebell, H., Zelder, O., and Pelzel, H.,** Fructose-induced depletion of liver adenine nucleotides in man, *Horm. Metab. Res.,* 3, 289, 1971.

84. **Itakura, M., Sabina, R. L., Heald, P. W., and Holmes, E. W.,** Basis for the control of purine biosynthesis by purine ribonucleotides, *J. Clin. Invest.,* 67, 994, 1981.

85. **Renold, A. E., Hastings, A. B., and Nesbett, F. B.,** Studies on carbohydrate metabolism in rat liver slices. III. Utilization of glucose and fructose by liver from normal and diabetic animals, *J. Biol. Chem.,* 209, 687, 1954.

86. **Thieden, H. I. D. and Lundquist, F.,** The influence of fructose and its metabolites on ethanol metabolism in vitro, *Biochem. J.,* 102, 177, 1967.

87. **Exton, J. H. and Park, C. R.,** Control of gluconeogenesis in liver. I. General features of gluconeogenesis in the perfused livers of rats, *J. Biol. Chem.,* 242, 2622, 1967.

88. **Seglin, P. O.,** Autoregulation of glycolysis, respiration, gluconeogenesis and glycogen synthesis in isolated parenchymal rat liver cells under aerobic and anaerobic conditions, *Biochim. Biophys. Acta,* 338, 317, 1974.

89. **Zakim, D., Pardini, R. S., Herman, R. H., and Sauberlich, H. E.,** Mechanism for the differential effects of high carbohydrate diets on lipogenesis in rat liver, *Biochim. Biophys. Acta,* 144, 242, 1967.

90. **Eggleston, L. V. and Woods, H. V.,** Activation of liver pyruvate kinase by fructose-1-phosphate, *FEBS Lett.,* 6, 43, 1970.

91. **Cook, G. C. and Jacobson, J.,** Individual variation in fructose metabolism in man, *Br. J. Nutr.,* 26, 187, 1971.

92. **Cook, G. C.,** Absorption and metabolism of D(-)fructose in man, *Am. J. Clin. Nutr.,* 24, 1302, 1971.

Chapter 9

INTERACTION WITH OTHER NUTRIENTS

I. INTRODUCTION

The effects of dietary components on metabolic and physiological processes are known to be significantly influenced by the presence or absence of other nutrients. There is evidence that fructose participates in many of these dietary interactions. If, as indicated by the material presented in Chapter 7, fructose is responsible for the major lipogenic effects of sucrose, then an interaction between saturated fat and fructose exists that produces larger increases in blood triglycerides of humans than does the combination of fructose and polyunsaturated fat.[1-4] In this chapter the interactions of dietary fructose with mineral status and bioavailability (especially that of copper), amino acid absorption, and ethanol metabolism will be described, possible mechanisms of action explored, and the significance to health and well-being discussed.

II. MINERALS

A. Copper

Some signs of copper deficiency in experimental animals include anemia; hypercholesterolemia; hypertriglyceridemia; hyperuricemia; abnormal electrocardiograms; cardiac hypertrophy, fibrosis, and rupture; weakening of arteries; pleural effusion; and often sudden death.[5] Increased blood cholesterol has also been reported in a human consuming a low-copper diet.[6] Relative or absolute deficiency of copper has been proposed to be a major factor in the etiology of human heart disease.[5-8]

In most studies in which the metabolic effects of copper deficiency in experimental animals have been reported, sucrose has been used as the source of dietary carbohydrate. The major rationale for its use is that sucrose is an extremely pure substance that would not be expected to provide any impurities containing trace minerals. Since both sucrose feeding and copper deficiency have been shown to produce some similar metabolic effects (i.e., increased blood triglycerides and uric acid), it was of interest to determine whether these dietary conditions were synergistic. A series of experiments was conducted, therefore, in order to determine the effects of different dietary carbohydrates on metabolic and physiological processes in rats fed diets either deficient or adequate in copper.[9-18] The animal model used was the weanling Sprague-Dawley rat. The major components of the basal diet (per kilogram diet) were 622 g carbohydrate (cornstarch, sucrose, fructose, or glucose), 200 g egg white, 90 g corn oil, 30 g cellulose, and 15 g rat vitamin mix. The mineral mixtures used (per kilogram diet) were either 40 g of Jones-Foster[19] with zinc added[9-13] or 35 g of AIN-76,[20] each with copper omitted.[14-18]

The most striking effect due to differences in the nature of the dietary carbohydrate in rats fed copper-deficient diets was on the incidence of their sudden death. Table 1 presents the mortality of rats fed copper-deficient diets containing starch, sucrose, fructose, or glucose. In the first study,[9] rats fed the copper-deficient diet containing sucrose began to die during the 5th week of feeding, and by the end of 8 weeks 6 out of 10 of these rats were dead. In contrast, none of the rats fed the copper-deficient diet containing cornstarch died during this time period. The higher incidence of mortality in the sucrose-fed rats could not be explained by differences in the copper content of the sucrose- or cornstarch-containing diets. In Study II, the comparative effects of cornstarch, sucrose, and fructose on the mortality of rats fed diets equally deficient in copper were determined.[10] After 7 weeks only 1 of 10 of the starch-

Table 1
MORTALITY OF RATS FED Cu-DEFICIENT DIETS CONTAINING CORNSTARCH, SUCROSE, FRUCTOSE, OR GLUCOSE AS THE CARBOHYDRATE SOURCE

Carbohydrate	Study I[a]		Study II[b]		Study III[c]	
	Dietary Cu (μg/g)	Mortality	Dietary Cu (μg/g)	Mortality	Dietary Cu (μg/g)	Mortality
Cornstarch	0.97	0/10	0.97	1/10	0.86	3/12
Sucrose	0.99	6/10	0.99	6/20	—	—
Fructose	—	—	1.01	7/20	0.92	10/15
Glucose	—	—	—	—	0.82	4/15

[a] Rats fed their respective diets for 8 weeks. Adapted from Fields, M., et al., *J. Nutr.*, 113, 1335, 1983.
[b] Rats fed their respective diets for 7 weeks. Adapted from Reiser, S., et al., *Am. J. Clin. Nutr.*, 38, 214, 1983.
[c] Rats fed their respective diets for 9 weeks. Adapted from Fields, M., et al., *Am. J. Clin. Nutr.*, 39, 289, 1984.

fed rats had died, while of 20 rats each fed sucrose or fructose, 6 and 7, respectively, had died. These results suggested that the fructose moiety of sucrose was responsible for the increased mortality. However, a nonspecific effect of all mono- or disaccharides on increasing mortality of rats fed copper-deficient diets could not be ruled out. Study III therefore was carried out, in which the effects of cornstarch, fructose, and glucose on the mortality of rats fed diets deficient in copper were observed.[11] In this study, copper levels in the diet were somewhat lower than in the previous two studies and the rats were fed for 9 weeks. This probably explains the higher mortality noted in the copper-deficient rats fed cornstarch in this study. The mortality rate was about the same in the glucose- (27%) as in the cornstarch-fed group (25%). In contrast, 10 of the 15 rats (67%) fed the copper-deficient diet containing fructose died, despite the copper content of this diet being about 12% higher than that of the glucose-containing diet. These results indicate that fructose or fructose-containing carbohydrates, as compared to glucose or glucose-containing carbohydrates, markedly increase the severity of copper deficiency in rats.

The results also may reconcile apparently inconsistent results of these effects of copper deficiency on mortality in rats. In studies where sucrose has been used,[21,22] numerous deaths were reported after 7 weeks. When starch was used, no deaths were reported.[23] The sudden death of the rats was apparently due to heart histopathology. Post-mortem examination of the rats revealed a large quantity of clotted blood in the pleural cavity. Hearts from the copper-deficient rats fed either sucrose or fructose, but not starch, were significantly hypertrophied as compared to their copper-supplemented controls (Table 2). The hypertrophied hearts from rats fed the sucrose or fructose diets deficient in copper had histopathological changes that included inflammation and fibrosis throughout the ventricular wall, large foci of necrosis, degeneration, incomplete repair of cardiac walls, and immature collagen in the areas of fibroblastic proliferation and repair.[9,10] Severe pericarditis was observed in some of these hearts. In contrast, the hearts from rats fed the copper-deficient diet containing starch (Studies I and II) were essentially normal in appearance and showed no abnormal histopathological changes.[9,10]

The comparative effects of starch and fructose on direct indices of copper status in rats fed diets deficient in or supplemented with copper are shown in Table 3. Serum copper concentration was drastically reduced by copper deficiency, regardless of the source of the

Table 2
HEART WEIGHTS OF RATS AFTER BEING FED FOR 7 WEEKS Cu-DEFICIENT OR Cu-SUPPLEMENTED DIETS CONTAINING CORNSTARCH, SUCROSE, OR FRUCTOSE

Dietary status	Heart weight (g/100 g body weight)
Starch fed, Cu deficient (n = 9)	0.45 ± 0.01[a]
Starch fed, Cu supplemented (n = 10)	0.40 ± 0.01[a]
Sucrose fed, Cu deficient (n = 10)	0.63 ± 0.02[b]
Sucrose fed, Cu supplemented (n = 10)	0.41 ± 0.01[a]
Fructose fed, Cu deficient (n = 10)	0.63 ± 0.04[b]
Fructose fed, Cu supplemented (n = 10)	0.38 ± 0.03[a]

Note: Values are expressed as mean ± SEM from the number of rats shown in parentheses. ANOVA revealed a significant effect of Cu, carbohydrate, and the interaction between them. Means with different superscript letters are significantly different from each other ($p < 0.05$) as determined by Duncan's multiple range test.

Adapted from Reiser, S., Ferretti, R. J., Fields, M., and Smith, J. C., Jr., *Am. J. Clin Nutr.*, 38, 214, 1983.

dietary carbohydrate. However, the concentration of serum copper was further reduced (about 47%) in copper-deficient rats fed fructose as compared to starch. Ceruloplasmin, a copper-containing protein found in blood, catalyzes the oxidation of the ferrous form of iron to the ferric form and thereby functions in hemopoiesis. Plasma ceruloplasmin activity is usually nondetectible after feeding a copper-deficient diet for 6 weeks and is therefore not a suitable parameter to differentiate degrees of severity of copper deficiency. Liver copper concentrations were significantly decreased in rats fed a copper-deficient diet as compared to the corresponding values from rats fed a copper-supplemented diet. In copper-deficient rats fed fructose, hepatic copper concentration was significantly decreased an additional 45% as compared to those fed starch. Superoxide dismutase (SOD), a copper-containing protein found in most tissues including liver, kidney, and erythrocytes, catalyzes the dismutation of the superoxide radical to hydrogen peroxide and oxygen, thereby participating in a process that protects cell membranes from peroxidation. Liver SOD activity was reduced by copper deficiency regardless of the nature of the dietary carbohydrate. However, feeding the fructose diet deficient in copper further reduced SOD activity by about 37% when compared to rats fed the copper-deficient diet containing starch. These results show that the feeding of fructose as compared to starch in diets deficient in copper produces a much more severe copper deficiency.

Copper-deficient rats fed fructose as compared to starch developed a relative anemia (Table 4). In individual animals the anemia was severe, with hemoglobin levels below 8 g/100 mℓ. The finding of anemia in copper deficiency is not surprising, since copper has been shown to participate in hemopoiesis[24] and copper-deficient animals show increased hemoglobin breakdown.[25] The differential effects of dietary carbohydrates on levels of hematocrit

Table 3
EFFECT OF Cu NUTRITURE AND NATURE OF DIETARY CARBOHYDRATE ON DIRECT INDICATORS OF Cu STATUS IN RATS

Dietary status	Serum Cu (μg/dℓ)	Plasma ceruloplasmin activity (U/ℓ)	Hepatic Cu (μg/g wet weight)	Hepatic SOD (U/g wet weight)
Starch fed, Cu deficient	4.9 ± 0.8[a]	< 0.1[a]	1.8 ± 0.1[a]	1038 ± 60[a]
Starch fed, Cu supplemented	107 ± 4[b]	98 ± 10[b]	4.2 ± 0.1[b]	1384 ± 100[b]
Fructose fed, Cu deficient	2.6 ± 0.1[c]	< 0.1[a]	1.0 ± 0.1[c]	660 ± 40[c]
Fructose fed, Cu supplemented	96 ± 5[b]	100 ± 8[b]	3.8 ± 0.1[d]	1330 ± 85[b]

Note: Each value represents the mean ± SEM from five rats. Means with different superscript letters within a column are significantly different from each other ($p < 0.05$) as determined by Duncan's multiple range test.

Values for serum Cu, plasma ceruloplasmin activity, and hepatic Cu were obtained from Fields, M., et al., *Am. J. Clin. Nutr.*, 39, 289, 1984. Values for hepatic SOD were obtained from Fields, M., et al., *Biol. Trace Elem. Res.*, 6, 379, 1984.

Table 4
EFFECT OF Cu NUTRITURE AND NATURE OF DIETARY CARBOHYDRATE ON INDICATORS OF HEMOPOIESIS IN RATS

Dietary status	Hematocrit (%)	Hemoglobin (g/100 mℓ)
Starch fed, Cu deficient (n = 9)	47.0 ± 1.3[a]	13.9 ± 0.2[a]
Starch fed, Cu supplemented (n = 10)	46.5 ± 0.9[a]	14.5 ± 0.3[a]
Fructose fed, Cu deficient (n = 10)	31.7 ± 2.9[b]	9.6 ± 0.9[b]
Fructose fed, Cu supplemented (n = 10)	46.1 ± 0.5[a]	13.7 ± 0.3[a]

Note: Each value represents the mean ± SEM from the number of rats shown in parentheses. Means with different superscript letters within a column are significantly different from each other ($p < 0.05$) as determined by Duncan's multiple range test.

Adapted from Reiser, S., Ferretti, R. J., Fields, M., and Smith, J. C., Jr., *Am. J. Clin. Nutr.*, 38, 214, 1983.

and homoglobin are probably a reflection of the much more severe copper deficiency in the rats fed fructose as compared to starch.

Total cholesterol and triglyceride levels in the blood were significantly increased by copper deficiency in both the starch- and fructose-fed rats (Table 5). However, the magnitude of increase in cholesterol and triglycerides was 79 and 31% greater ($p < 0.05$), respectively, in rats fed fructose than in those fed starch. No correlation was found between the level of blood cholesterol and the extent of cardiac damage. Except for one rat, no cardiac damage was found in copper-deficient rats fed starch, despite cholesterol levels significantly higher than those of the copper-supplemented controls. These findings argue against a role of elevated cholesterol levels in the etiology of the cardiac pathology observed in fructose-fed rats.

Table 5
EFFECT OF Cu NUTRITURE AND NATURE OF DIETARY CARBOHYDRATE ON BLOOD LIPIDS IN RATS

Dietary status	Total cholesterol (mg/100 mℓ)	Triglycerides (mg/100 mℓ)
Starch fed, Cu deficient ($n = 9$)	113 ± 8[a]	49 ± 5[a]
Starch fed, Cu supplemented ($n = 10$)	70 ± 5[b]	37 ± 2[b]
Fructose fed, Cu deficient ($n = 10$)	202 ± 30[c]	64 ± 6[c]
Fructose fed, Cu supplemented ($n = 10$)	71 ± 3[b]	34 ± 3[b]

Note: Each value represents the mean ± SEM from the number of rats shown in parentheses. Means with different superscript letters within a column are significantly different from each other ($p < 0.05$) as determined by Duncan's multiple range test.

Adapted from Reiser, S., Ferretti, R. J., Fields, M., and Smith, J. C., Jr., *Am. J. Clin. Nutr.*, 38, 214, 1983.

Table 6
EFFECT OF Cu NUTRITURE AND NATURE OF DIETARY CARBOHYDRATE ON MEMBRANE PEROXIDATION AND HEPATIC ATP LEVELS IN RATS

Dietary status	Mitochondrial peroxidation (nmol malonaldehyde/μg protein/hr at 37°C) Liver	Heart	Hepatic ATP (μmol/g liver wet weight)
Starch fed, Cu deficient	4.1 ± 0.4[a]	4.4 ± 0.3[a]	0.96 ± 0.12[a]
Starch fed, Cu supplemented	3.6 ± 0.3[a]	3.4 ± 0.6[a]	1.04 ± 0.06[a]
Fructose fed, Cu deficient	6.3 ± 0.2[b]	5.8 ± 0.3[b]	0.55 ± 0.12[b]
Fructose fed, Cu supplemented	3.3 ± 0.3[a]	4.5 ± 0.4[a]	1.22 ± 0.10[a]

Note: Each value represents the mean ± SEM from five rats. Means with different superscript letters within a column are significantly different from each other ($p < 0.05$) as determined by Duncan's multiple range test.

Adapted from Fields, M., Ferretti, R. J., Smith, J. C., and Reiser, S., *Biol. Trace Elem. Res.*, 6, 379, 1984.

Peroxidation values of mitochondrial preparations obtained from both the liver and heart, as well as hepatic ATP levels from copper-deficient and copper-supplemented rats fed either starch or fructose, are shown in Table 6. Rats fed the starch diet deficient in copper showed no increase in tissue peroxidation or decrease in ATP levels when compared to copper-supplemented controls. In contrast, copper deficiency significantly increased hepatic and myocardial mitochondrial peroxidation and decreased ATP levels in rats fed fructose as compared to their copper-supplemented controls. SOD is involved in the prevention of tissue peroxidation,[26] and the reduction of this enzyme has been associated with enhanced cell damage.[27,28] The reduced levels of SOD found in fructose-fed, copper-deficient rats are

Table 7

EFFECT OF THE REPLACEMENT OF FRUCTOSE BY STARCH IN Cu-DEFICIENT DIETS FED TO RATS ON VARIOUS METABOLIC PARAMETERS

Dietary conditions	Hematocrit (%)	Hemoglobin (g/100 mℓ)	Cholesterol (mg/100 mℓ)	Triglycerides (mg/100 mℓ)	Hepatic SOD (U/g wet weight)
Rats fed Cu-deficient diets containing fructose for 11 weeks	39.2 ± 0.6[a]	12.9 ± 0.2[a]	147 ± 9[a]	65 ± 11[a]	660 ± 40[a]
Rats fed Cu-deficient diets containing fructose for 5 weeks and then fed Cu-deficient diets containing starch for an additional 6 weeks	45.0 ± 1.3[b]	13.7 ± 0.1[b]	87 ± 8[b]	37 ± 5[b]	972 ± 62[b]

Note: Each value represents the mean ± SEM from five rats. Means with different superscript letters within a column are significantly different from each other ($p < 0.05$) as determined by ANOVA and Duncan's multiple range test.

Values for hematocrit, hemoglobin, cholesterol, and triglycerides were obtained from Fields, M., et al., *Proc. Soc. Exp. Biol. Med.*, 178, 362, 1985. Values for hepatic SOD were obtained from Fields, M., et al., *Biol. Trace Elem. Res.*, 6, 379, 1984.

compatible with these results. It is possible that peroxidation of cardiac membranes contributes to the heart histopathology observed in rats fed the copper-deficient diet containing fructose. Lipid peroxidation of mitochondria, with resultant uncoupling of oxidative phosphorylation, could also explain the reduced levels of ATP found in the copper-deficient rats fed fructose.

The mechanisms by which the feeding of fructose or sucrose as compared to glucose-based carbohydrates exacerbates the signs associated with copper deficiency are as yet unknown. The undesirable efects of fructose feeding on several metabolites indicative of copper deficiency are apparently reversible. When rats were fed a copper-deficient diet containing fructose for 5 weeks and then fed a copper-deficient diet containing starch for an additional 6 weeks, hematocrit, hemoglobin, and hepatic SOD levels were significantly higher and cholesterol and triglyceride values significantly lower when compared to the corresponding levels of these parameters in rats fed a copper-containing diet with fructose for 11 weeks (Table 7).

These results emphasize the importance of interactions that may occur between dietary components and the resultant changes in the bioavailability of nutrients. Two likely sites of interaction between fructose and copper would be the liver and the intestine. Since the metabolism of fructose in rats and man occurs mainly in the liver, an interaction between fructose or any one of its metabolites and copper that affects the utilization of copper may occur in this tissue. Events occurring in the small intestine also may be the focus of this interaction. Carbohydrates have been reported to form chelates with minerals in the intestine which may alter the absorption of inorganic elements.[29-32] More specifically, it has been reported that copper absorption was reduced in rats receiving fructose rather than starch as the dietary carbohydrate, but only when the diet contained excessive levels of iron.[33]

In order to investigate the role of intestinal absorption on the interaction between fructose and copper status, copper retention in the GI tract was determined following the intragastric intubation of [67]Cu-containing meals into rats previously fed fructose or starch diets deficient in or supplemented with copper for 5 weeks.[17] All rats fed the fructose diets (deficient in or supplemented with copper) were intubated with a slurry of the copper-supplemented fructose diet, and rats fed the starch diet (deficient in or supplemented with copper) were

Table 8

**GI RETENTION (% OF RADIOACTIVITY
ADMINISTERED) OF ^{67}Cu AT VARIOUS TIMES
AFTER INTRAGASTRIC ADMINISTRATION TO Cu-
DEFICIENT OR Cu-SUPPLEMENTED RATS FED
FRUCTOSE OR STARCH**

Time after administra- tion (hr)	Cu deficient		Cu supplemented	
	Fructose	Starch	Fructose	Starch
8	57	34	60	57
24	60	43	48	37
48	29[a]	18	22	22
96	21[a]	6	13	12

Note: Each value represents the mean for five rats.

[a] Significantly greater ($p < 0.05$) than corresponding starch-fed value as determined by orthogonal contrasts.

Adapted from Fields, M., Holbrook, J., Scholfield, D., Smith, J. C., Jr., Reiser, S., and Los Alamos Medical Research Group, *J. Nutr.*, 116, 625, 1986.

given a slurry of the copper-supplemented starch diet. Table 8 shows the GI retention of ^{67}Cu at various times after the intragastric administration of the meals. Copper-deficient rats fed fructose retained more radioactivity in the GI contents than did those fed starch. The retention of ^{67}Cu by the carcasses of copper-deficient rats fed fructose was generally reduced as compared to those fed starch.[17] No differences in the GI retention of ^{67}Cu were observed on copper-supplemented rats as a function of dietary carbohydrate. These results suggest that intestinal events are at least partly responsible for the interaction between fructose and copper status in rats fed copper-deficient diets.

Diets consumed by people living in the U.S. have relatively high levels of fructose-containing carbohydrates and fructose.[34-36] The intake of copper in the U.S. appears to be well below the daily level of 2 mg considered to be adequate by the Food and Nutrition Board of the National Academy of Sciences.[37-39] If the same type of interaction between dietary carbohydrate and copper status found in rats also occurs in humans, then the high intake of sucrose and/or fructose may lower copper levels that are already marginal due to inadequate copper intake. A human study was therefore initiated to determine the effect of feeding fructose as compared to starch in diets providing an average of 1 mg of copper per day on indices of copper status such as serum copper, serum ceruloplasmin activity, and erythrocyte SOD activity.[40] The diets contained 49% of calories as carbohydrate, 37.5% as fat (P/S ratio of 0.33), and 13.5% as protein, and provided a daily average of 5.6 g of crude fiber and 583 mg of cholesterol. The only difference in the diets was that 40% of the carbohydrate calories were from either fructose or cornstarch. The diets were to be fed for 7 weeks each in a crossover pattern. Due to health-related problems experienced by four of the subjects, the study was terminated after 11 weeks (7 weeks of Period I, 4 weeks of Period II). Table 9 presents the effect of fructose as compared to cornstarch on indices of copper status after the diets were fed for 4 weeks each. Serum copper concentration was neither significantly affected by diet or period, nor was there an interaction between diet and period. Ceruloplasmin activity was not significantly affected by diet, but showed a significant period effect, activity being higher in Period II than Period I. The activity of SOD in the erythrocytes was significantly affected by diet and a significant interaction was

Table 9

**SERUM Cu, CERULOPLASMIN ACTIVITY, AND ERYTHROCYTE SOD
ACTIVITY AFTER SUBJECTS WERE FED FRUCTOSE AND STARCH
DIETS FOR 4 WEEKS EACH**

Diet	Serum Cu[a] (μg/100 mℓ)		Serum ceruloplasmin activity[b] (U/ℓ)		Erythrocyte SOD activity[c] (U/10⁹ erythrocytes)	
	Period I	Period II	Period I	Period II	Period I	Period II
Fructose	111 ± 6 (9)	109 ± 6 (11)	94 ± 10 (9)	111 ± 11 (11)	34.4 ± 1.4[a,b] (9)	31.6 ± 1.4[b] (11)
Starch	110 ± 6 (11)	113 ± 6 (9)	102 ± 11 (11)	116 ± 11 (9)	35.7 ± 1.5[a,b] (11)	38.6 ± 1.5[a] (9)

Note: Each value represents the mean ± SEM from the number of subjects shown in parentheses. Values in a column not sharing a common superscript letter are significantly different ($p < 0.05$) according to Duncan's multiple range test. The two groups of subjects initially showed no significant differences in SOD activity ($p = 0.80$, unpaired difference t-test).

[a]   ANOVA: Diet — $p = 0.58$; period — $p = 0.68$; interaction — $p = 0.68$.
[b]   ANOVA: Diet — $p = 0.25$; period — $p < 0.03$; interaction — $p = 0.88$.
[c]   ANOVA: Diet — $p < 0.005$; period — $p = 0.81$; interaction — $p < 0.05$.

Adapted from Reiser, S., Smith, J. C., Jr., Mertz, W., Holbrook, J. T., Scholfield, D. J., Powell, A. S., Canfield, W. K., and Canary, J. J., *Am. J. Clin. Nutr.*, 42, 242, 1985.

Table 10

**SERUM Cu, CERULOPLASMIN ACTIVITY, AND ERYTHROCYTE SOD
ACTIVITY BEFORE (11TH WEEK) AND AFTER (14TH WEEK) REPLETION
WITH 3 mg Cu PER DAY FOR 3 WEEKS**

Index of Cu status	Previously fed fructose ($n = 11$)		Previously fed starch ($n = 9$)	
	Before repletion	After repletion	Before repletion	After repletion
Serum Cu (μg/100 mℓ)	109 ± 6	101 ± 4	113 ± 6	114 ± 4
Serum ceruloplasmin activity (U/ℓ)	111 ± 11	108 ± 4	116 ± 11	100 ± 7
Erythrocyte SOD activity (U/ 10⁹ erythrocytes)	31.6 ± 1.4	40.4 ± 2.0[a]	38.6 ± 1.5	33.4 ± 1.2

Note: Each value represents the mean ± SEM from the number of subjects shown in parentheses.

[a]   Significantly greater ($p < 0.05$) than corresponding value before repletion according to a paired difference t-test.

Adapted from Reiser, S., Smith, J. C., Jr., Mertz, W., Holbrook, J. T., Scholfield, D. J., Powell, A. S., Canfield, W. K., and Canary, J. J., *Am. J. Clin. Nutr.*, 42, 242, 1985.

found between diet and period. SOD activity was 18% lower ($p < 0.05$) in subjects fed fructose as compared to starch only after Period II or after the subjects had consumed the low-copper diet for 11 weeks. There was no significant difference in SOD activity between the two subject groups before the diets were fed ($p = 0.80$, unpaired difference t-test).

After the 11th week of the study all subjects were fed the starch-containing diet supplemented with 3 mg of copper per day for 3 weeks.[40] Table 10 presents a comparison of the levels of serum copper, ceruloplasmin activity, and erythrocyte SOD activity before and after

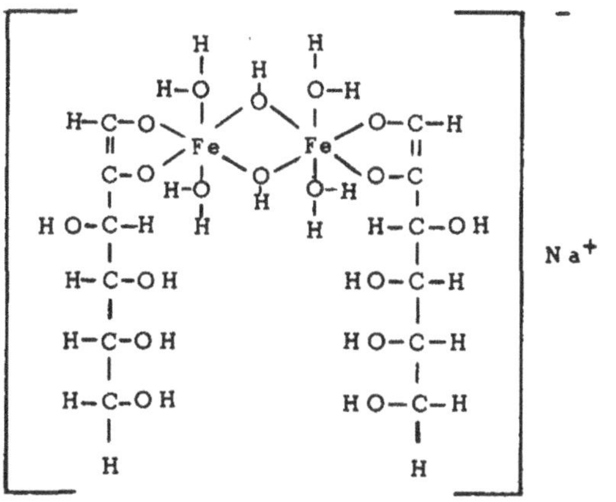

FIGURE 1. Proposed hypothetical structure for an iron-fructose complex. (Adapted from Charley, P. J., Bibudhendra, S., Stitt, C. F., and Saltman, P., *Biochim. Biophys. Acta*, 69, 313, 1963.)

repletion with copper. Serum copper and ceruloplasmin activity were not significantly different before or after copper repletion. In contrast, SOD activity was significantly higher ($p < 0.04$) after than before repletion, but only when the subjects had previously been fed the fructose diet.

The measurements of serum copper concentration, ceruloplasmin activity, and SOD activity can provide useful indices of copper status in humans. On the basis of a previous study,[6] SOD appears to be the most sensitive indicator of copper status in humans. These results suggest that the type of dietary carbohydrate fed can affect differentially a sensitive indicator of copper status in humans.

B. Iron

Of the common dietary sugars, fructose appears to have the greatest capacity to form complexes with iron when both are present in aqueous solutions.[41] An iron-fructose complex has been isolated and purified, and elemental analysis indicates the complex contains two atoms of iron, two molecules of fructose, and one atom of sodium. A hypothetical model for this complex is shown in Figure 1. The ability of fructose to form a chelate with iron has been attributed to the dihydroxyacetone structure found in the open chain of the fructose molecule, but not in any aldohexose or disaccharide.[42] The finding that sorbose and tagatose, but not glucose, galactose, mannose, sucrose, lactose, or maltose, form chelates with iron is consistent with this structural prerequisite.

It appears that the iron-fructose complex facilitates the absorption and retention of iron to a greater extent than do other iron compounds. Figure 2 presents the results of a study in which whole-body retention of iron by guinea pigs was determined 10 days after the oral administration of radioactive iron chelated to fructose, citrate, nitrilotetracetic acid (NTA), ethylene-diaminetetracetic acid (EDTA), or gluconate.[43] The retention of iron from $^{59}FeSO_4$ was used as a basis of comparison. The retention of iron from fructose was anywhere from two to three times higher than from the other iron-containing substances. Previous studies also have indicated that the iron-fructose complex is effectively absorbed by humans, rats, rabbits, and swine.[43]

The relevance of these findings to iron nutriture remains questionable, being dependent upon formation of an iron-fructose complex under normal dietary conditions. Serum iron was not reported to be significantly different 6 weeks after rats were fed diets either adequate or high in iron and containing 75% by weight of cornstarch, sucrose, glucose, or fructose.[33]

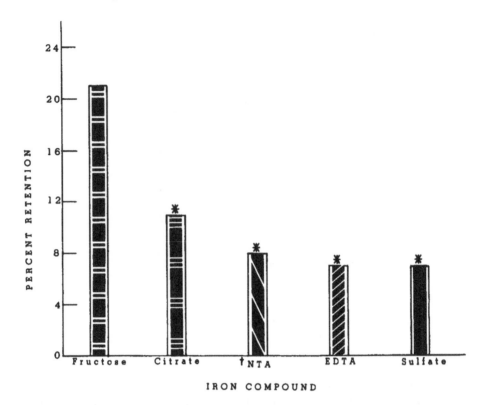

FIGURE 2. Retention of 28 μg of iron from each compound 10 days after being given as a single 1-mℓ oral dose. Each value represents the mean from six guinea pigs. (*) Significantly lower than retention obtained with fructose ($p < 0.05$). (†) Nitrilotetracetic acid. (Adapted from Bates, G. W., Boyer, J., Hegenauer, J. C., and Saltman, P., *Am. J. Clin. Nutr.*, 25, 983, 1972.)

Similarly, the hepatic iron content of rats fed 72% fructose for 5 to 9 weeks was not higher than that of rats fed 72% cornstarch[11-17] or glucose.[22] However, the addition of an iron-fructose complex to certain foods or its use as an iron supplement may be a useful and pleasurable way by which to meet effectively the requirements for iron.

C. Selenium

There has been increasing evidence recently that adequate selenium nutriture is essential for human health and well-being. The incidence of Keshan disease, a cardiomyopathy endemic in certain regions of the People's Republic of China, was dramatically reduced after an intervention trial with sodium selenite.[44,45] Epidemiological studies also suggest a relationship between low intake of selenium and cardiovascular disease.[46] A relationship also appears to exist between increased incidence of colon and mammary cancer and low levels of environmental selenium.[47-49] From these reports it is evident that any nutritional interaction that produces a decrease in the bioavailability or metabolic utilization of selenium could have a serious effect on human health.

An interaction may exist between copper and selenium status. The feeding to rats of diets based on evaporated milk deficient in copper was reported to decrease the activities of both liver SOD and the selenoenzyme, glutathione peroxidase.[50] The levels of glutathione peroxidase in various tissues of the rat have been found to be a good measure of selenium status.[51] Copper is utilized in metabolic pathways that are closely related to those utilized by selenium.[52] The intermediates of oxygen metabolism, the superoxide radical ($O_2^{.-}$) and hydrogen peroxide (H_2O_2), are scavenged by SOD, catalase, and glutathione peroxidase, providing protection against tissue peroxidation and membrane damage.[53] Since fructose has

Table 11
EFFECT OF Cu STATUS AND DIETARY CARBOHYDRATE ON THE ACTIVITY OF HEPATIC GLUTATHIONE PEROXIDASE IN RATS

Dietary status	Hepatic glutathione peroxidase activity (μmol NADPH oxidized/mg protein/min)
Cornstarch fed, Cu deficient	1.00 ± 0.15^a
Cornstarch fed, Cu supplemented	1.09 ± 0.15^a
Fructose fed, Cu deficient	0.50 ± 0.10^b
Fructose fed, Cu supplemented	0.75 ± 0.02^b
Glucose fed, Cu deficient	0.95 ± 0.11^a
Glucose fed, Cu supplemented	0.90 ± 0.04^a

Note: Each value represents the mean $\pm$ SEM from five rats. Data were analyzed by 3 $\times$ 2 analysis of variance. Effects and interactions were significant at the ($p < 0.05$ level. Values within a column with different superscript letters are significantly different from each other ($p < 0.05$) according to Duncan's multiple range test. ANOVA: Copper — significant; carbohydrate — significant; interaction — significant.

Adapted from Fields, M., Ferretti, R. J., Smith, J. C., and Reiser, S., *Biol. Trace Elem. Res.*, 6, 379, 1984.

been shown to affect copper status in rats negatively,[9-18] it was of interest to determine the effects on selenium status in rats fed copper-deficient or -supplemented diets adequate in selenium and containing cornstarch, fructose, or glucose as the source of dietary carbohydrate.[14,15] The effect of copper deficiency and the nature of the dietary carbohydrate on hepatic glutathione peroxidase activity is shown in Table 11. Glutathione peroxidase activity was significantly reduced by 50% in copper-deficient rats fed fructose. In contrast, copper-deficient rats fed cornstarch or glucose showed no significant difference in enzyme activity as compared to comparably fed copper-supplemented rats. The copper deficiency in the fructose-fed rats which decreased glutathione peroxidase activity as well as SOD activity (Table 3) also accentuated the peroxidation of tissue membranes (Table 6). Interestingly, the feeding of fructose as compared to cornstarch or glucose also significantly decreased glutathione peroxidase activity by about 25% in copper-supplemented rats. These results are consistent with a relationship between copper and selenium status and also indicate a direct effect of fructose feeding on selenium status that is independent of copper status. In this regard it has been reported that a gradual decrease of glutathione peroxidase activity occurred in the blood and liver of rats fed sucrose or fructose as compared to starch.[54]

III. AMINO ACIDS

It appears that monosaccharides derived from dietary sources can affect differentially the absorption of amino acids by rat intestine. Table 12 summarizes some of the studies reporting the effects of glucose, galactose, and fructose on the transport of amino acids by epithelial cells isolated from rat intestine.[55-57] The transport of neutral amino acids such as leucine, isoleucine, valine, phenylalanine, tryptophan, and histidine was increased 24 to 57% by the presence of fructose in the incubation media. The fructose stimulation does not depend on the use of isolated cells as the intestinal preparation, since studies using intact intestine also have shown a stimulation of neutral amino acid transport by fructose.[58,59] In contrast to fructose, glucose and galactose generally inhibited the transport of these neutral amino acids. These findings support the existence of independent intestinal pathways for the transport of

Table 12
EFFECT OF GLUCOSE, GALACTOSE, AND FRUCTOSE ON THE TRANSPORT OF AMINO ACIDS BY EPITHELIAL CELLS ISOLATED FROM RAT INTESTINE

Conditions	Amino acid	Effect of monosaccharides on amino acid transport (% of control activity)[a]			Ref.
		Glucose	Galactose	Fructose	
Measurement of intracellular accumulation of 1 mM of the amino acid after a 15-min incubation with 25 mM of each monosaccharide (n = 11—12)	L-Leucine	80.7[b]	49.2[b]	141.5[b]	55
Measurement of intracellular accumulation of 1 mM of the amino acid after a 5-min incubation with 10 mM of the indicated monosaccharide (n = 7)	L-Leucine	42.4[b]		134.4[b]	56
	L-Lysine	95.0		102.0	
Same as in preceding study (Reference 56) (n = at least 7)	L-Leucine	69.7[b]		132.0[b]	57
	L-Isoleucine	100.0		157.1[b]	
	L-Valine	76.8[b]		126.8[b]	
	L-Phenylalanine	66.0[b]		127.9[b]	
	L-Tryptophan	99.4		127.3[b]	
	L-Histidine	83.7		124.0	
	L-Lysine	86.2[b]		95.3	
	L-Ornithine	92.8		89.9	

[a] Percentage of control activity obtained by dividing the average intracellular accumulation of the amino acid in the presence of the monosaccharide by the average intracellular accumulation of the amino acid in the absence of the monosaccharide (control) and multiplying by 100.

[b] Significantly different from control ($p < 0.05$) as determined by a paired difference t-test.

glucose and fructose. Fructose failed to stimulate the transport of lysine and ornithine, which utilize the basic amino acid transport pathway.

The finding that intracellular was as effective as extracellular fructose in stimulating the uptake of leucine raised the possibility that the stimulation was mediated by an exchange transport between intracellular fructose and extracellular neutral amino acids. However, experiments designed to detect the presence of exchange transport produced results incompatible with this type of interaction.[57] These negative results can be summarized as follows:

1. No concurrent inhibition of fructose transport occurred under conditions where stimulation of neutral amino acid transport was found.
2. Neutral amino acids whose transport was stimulated by fructose were unable to accelerate the movement of fructose out of fructose-loaded epithelial cells.
3. Fructose showed no tendency to participate in exchange transport.
4. The component of leucine transport stimulated by intracellular fructose was still dependent on metabolic energy and the presence of a Na$^+$ gradient.

There appears to be a major quantitative difference in the Na$^+$ dependence of the neutral and basic amino acid transport systems. The transport of basic amino acids is much less dependent on Na$^+$ than that of neutral amino acids.[60,61] It is therefore possible that fructose stimulation is related to the Na$^+$ dependence of a transport system. However, the finding that fructose did not stimulate the transport of glucose, galactose, or α-methylglucose[57] is not consistent with such a relationship.

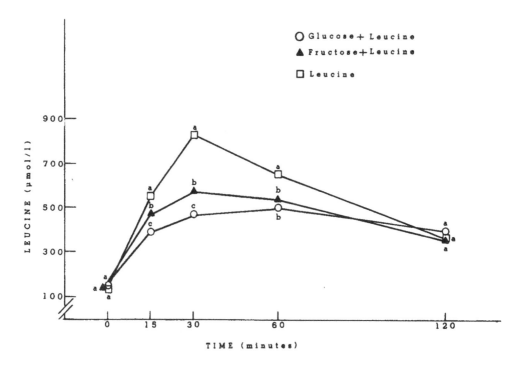

FIGURE 3. Plasma leucine responses after humans consumed 300-mℓ drinks containing 5 g leucine per 70 kg body weight without or with ten times the molar concentration of either glucose or fructose. Each point represents the mean from 23 subjects. Leucine values at each time point not sharing a common superscript letter are significantly different ($p < 0.05$) according to Duncan's multiple range test. (Adapted from Reiser, S., Scholfield, D., Trout, D., Wilson, A., and Aparicio, P., *Nutr. Rep. Int.*, 30, 151, 1984.)

A difference in the effect of glucose and fructose on the absorption of leucine also has been observed in humans.[62] A group of 23 subjects ranging in age from 25 to 57 years were given 300-mℓ drinks containing 5 g of leucine per 70 kg body weight without or with ten times the molar concentration of either fructose or glucose (69 g/70 kg body weight) in a crossover design separated by at least 3 weeks. Blood was drawn before and 15, 30, 60, and 120 min after the drinks were consumed, and plasma was analyzed for leucine. Figure 3 shows that plasma leucine levels were significantly higher 15, 30, and 60 min after the subjects drank the leucine without the sugars than with the sugars. However, leucine levels of subjects given the leucine drink with fructose were 27 and 20% higher than with glucose at 15 and 30 min, respectively. The summation of the area under the leucine response curves 15 to 120 min after consuming the leucine-fructose drink was 14% higher ($p < 0.05$) than after the leucine-glucose drink.

Differences in the gastric emptying of the glucose and fructose drinks may explain their comparative effects on leucine absorption. It has been shown that glucose is more effective than an equal amount of fructose in slowing gastric emptying in humans under controlled feeding conditions.[63] In addition, the greater elevation of blood glucose after the feeding of glucose as compared to fructose would also contribute to the slower gastric emptying of the glucose drink, since an elevated blood glucose level has been reported to slow gastric emptying.[64]

These results indicate that leucine absorption in humans is more rapid when fructose as compared to glucose is consumed concurrently. Whether this interaction is of nutritional significance, especially in societies where protein malnutrition is a problem, remains to be determined.

IV. ETHANOL

An estimated 61% of American adults consume alcoholic beverages on a regular basis,[65] and each year over one half of all fatal motor vehicle accidents are attributed to alcohol consumption.[66] The blood alcohol concentration reflects the concentration of ethanol in body tissues and is directly related to the level of functional impairment due to alcohol. The ability to increase the rate of removal of ethanol from the tissues would have far-reaching social and medical implications. Many substances such as hormones, vitamins, amino acids, substrates of carbohydrate metabolism, and mono- and disaccharides have been tested in pursuit of a method to increase the rate of ethanol metabolism.[67-71] Of the substances tested, none but fructose has been reported to be of benefit in repeated trials with human subjects. Sucrose also has been shown to have a slight effect in increasing ethanol metabolism,[71,72] presumably due to its fructose moiety. In this section, results from studies in which the effect of fructose on ethanol metabolism has been investigated will be presented and mechanisms proposed to explain this interaction will be discussed.

Although the effect of fructose on ethanol metabolism in humans has been investigated for almost 50 years, no clear consensus on the magnitude of the effect or the responsible mechanism has yet been achieved. As early as 1937 it was reported that the disappearance of ethanol in the expired air occurred more quickly after the oral intake of fructose as compared to glucose.[73,74] In the 1950s there were several reports which claimed that fructose had an accelerating effect on ethanol metabolism.[75-77] For example, a 50% increase in the maximum velocity of ethanol oxidation was found with fructose as compared to ethanol alone.[77]

More recent studies have reported lower blood alcohol levels when fructose was given with ethanol than when ethanol was given either with a placebo or alone. The level of blood alcohol and the rate of ethanol clearance were compared in 11 subjects given ethanol plus noncaloric placebo or ethanol plus 97.3 g of fructose on two separate occasions.[67] The mean blood alcohol level was lower with fructose at $1\frac{1}{2}$ and 2 hr ($p < 0.02$), but the rate of alcohol disappearance was similar in the two groups from 2 to 3 hr. Similarly, significantly lower blood alcohol levels were found when ethanol (0.798 g/kg body weight) was preceded by 125 g of fructose as compared to ethanol preceded by placebo.[71] Urine output also was monitored and found to be decreased by fructose. It was postulated, therefore, that differences due to the iso-osmotic holding of ethanol and water in the intestine by fructose were responsible for the fructose-induced decrease in blood ethanol. The rate of ethanol disappearance was reported to be increased significantly ($p < 0.05$) when 9 subjects were given 135 g of fructose 3 hr after a load dose of 94.7 g of ethanol as compared to 13 controls given ethanol alone.[78] A significant decrease ($p < 0.05$) in the area under the blood alcohol curve was found when 10 subjects consumed ethanol plus 60 g of fructose in comparison to the same amount of ethanol alone.[79]

Some studies have compared the effects of fructose on ethanol metabolism to those of other compounds. The rate of clearance of 0.5 g/kg body weight of ethanol with and without the concurrent consumption of 30 g of glucose, galactose, fructose, or sucrose was compared in 12 men using a crossover design.[68] Consumption of glucose and galactose produced no change in the rate of ethanol metabolism as compared to ethanol alone. In contrast, the consumption of fructose was reported to cause a significant 18 to 40% increase in the rate of ethanol clearance over ethanol alone. Sucrose consumption was reported to cause a very slight increase in ethanol clearance (levels of significance not reported). In another study,[80] 52 alcoholic subjects were given 0.78 g of ethanol per kilogram body weight followed by 0.5 g/kg body weight of fructose, glucose, sucrose, or alanine at 60 min. Blood alcohol was reported to be lower at 90 and 120 min with fructose than with the other compounds tested. A 100% increase in the clearance rate of ethanol was observed with fructose as

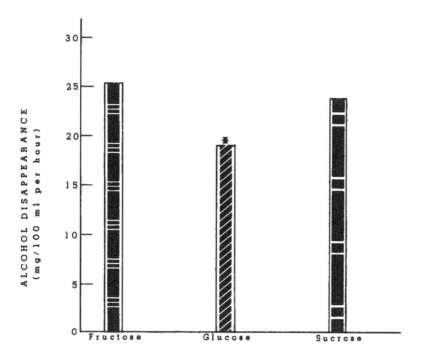

FIGURE 4. Effect of the previous consumption (per kilogram body weight) of 2 g of fructose, 2 g of glucose, or 4 g of sucrose on the rate of removal of intravenously administered alcohol (1 g per kilogram body weight) from the blood of alcoholic men. Each value represents the mean from eight subjects. (*) Significantly lower ($p < 0.05$) than corresponding fructose or sucrose disappearance rates as determined by student's t-test. (Adapted from Soterakis, J. and Iber, F. L., *Am. J. Clin. Nutr.*, 28, 254, 1975.)

compared to alanine. Employing a somewhat different experimental approach, the effect of oral administration (per kilogram body weight) of 2 g of glucose, 2 g of fructose, or 4 g of sucrose on the disappearance rate of intravenously administered alcohol (1 g/kg body weight) was determined in eight alcoholic men.[72] On the basis of blood alcohol determinations obtained hourly after alcohol administration, the rate of alcohol disappearance was significantly greater when either fructose or sucrose rather than glucose was consumed (Figure 4).

Other studies relating to effects on ethanol disappearance have been conducted in which the sugars have been given by infusion in order to circumvent possible differences caused by intestinal absorption. A group of 12 alcoholic subjects were given 94.7 g of ethanol orally while they were being infused with 750 mℓ of 10% glucose, 10% fructose, or normal saline.[81] Blood alcohol levels were lower ($p < 0.01$) while fructose as compared to saline was being infused. There was no significant difference in blood alcohol levels between glucose and saline infusion. Results of the glucose and fructose trials were not compared statistically. Intravenous fructose also has been tested as an aid in lowering the blood alcohol levels in acutely intoxicated subjects.[82] The rate of decrease of ethanol in the blood of these subjects was significantly higher ($p < 0.05$) after infusion with 200 g of fructose than after saline. Other studies also have confirmed the acceleration of ethanol clearance in the blood by intravenously administered fructose.[83,84]

Not every study has reported a specific beneficial effect of fructose in lowering blood alcohol or increasing the rate of ethanol oxidation. Subjects given 60 g of ethanol with 39 or 78 g of fructose were found to have inconsistent alterations in blood alcohol levels and no change in the rate of ethanol metabolism as compared to ethanol alone.[85] In another study, 12 subjects were given 1 g/kg body weight of ethanol either with a placebo or with

1.2 g/kg body weight of glucose or fructose.[86] Although blood alcohol levels were significantly lower ($p < 0.05$) at 100 min with glucose and at 100, 160, and 220 min with fructose, compared to ethanol with placebo, blood alcohol levels for the ethanol and fructose group did not differ from those of the ethanol and glucose group. There was also no significant difference in the slope of the ethanol elimination curve among the three groups. Similar levels of blood alcohol, rates of ethanol disappearance, and subjective and objective signs of intoxication were reported after 109 male subjects were given 1.75 g/kg body weight of ethanol without or with 1 g/kg body weight of either glucose or fructose.[87]

Taken as a whole, these studies indicate that fructose given together with ethanol as compared to ethanol given alone produces lower levels of blood alcohol and increases the rate of ethanol clearance. Some of the studies suggest that this is a unique property of fructose or the fructose moiety of sucrose rather than a nonspecific effect shared by all sugars. The conditions required for the optional manifestation of this effect of fructose, its specificity, and its clinical implications are areas in need of further investigation.

Many mechanisms have been proposed to explain the effect of fructose on ethanol metabolism. In general, it is believed that fructose in some way speeds the reoxidation of NADH, which is generated in the oxidation of ethanol to acetate by alcohol dehydrogenase and aldehyde dehydrogenase. The most commonly cited mechanisms include: (1) increased reoxidation of NADH in the cytosol by the reduction of glyceraldehyde to glycerol or the reduction of fructose to sorbitol, (2) increased reoxidation of NADH via the malic enzyme shuttle, and (3) increased reoxidation of NADH via oxidative phosphorylation in the mitochondria.

Glyceraldehyde resulting from fructose metabolism may be converted to glycerate by aldehyde dehydrogenase. However, in the presence of ethanol this pathway may be inhibited by acetaldehyde, which has a high affinity for aldehyde dehydrogenase, and by the reduced ratio of NAD^+ to NADH in the liver. Glyceraldehyde may instead be reduced to glycerol by the alcohol dehydrogenase-NADH complex. If this occurs, dissociation of alcohol dehydrogenase and NADH, the rate-limiting step in the reoxidation of the complex, would be by-passed and alcohol dehydrogenase-NADH would be reoxidized directly.[88] This mechanism is supported by the finding that glycerol and glycerol-3-phosphate accumulated during fructose-stimulated ethanol oxidation in isolated rat liver cells.[89] However, on the basis of a study using perfused rat liver, it was estimated that this pathway could account for less than 20% of the increase in rate of ethanol oxidation due to fructose.[90]

The reduction of fructose to sorbitol is normally of little importance in fructose metabolism. However, when fructose and ethanol are consumed together, higher blood concentrations of sorbitol are seen than during the ingestion of fructose alone.[80] In isolated rat liver cells incubated with ethanol and fructose, sorbitol was reported to accumulate in substantial quantities.[89] Another study using perfused pig liver has dismissed sorbitol formation and the accompanying reoxidation of NADH as insignificant in the stimulation of ethanol oxidation by fructose.[91]

It has been proposed that a mechanism involving the "malic enzyme shuttle" is responsible for the twofold increase in ethanol oxidation by fructose in perfused pig liver. Figure 5 shows the metabolic pathways involved. The major feature of this sequence is the reoxidation of NADH generated by the alcohol dehydrogenase reaction due to the conversion of oxaloacetate to malate by malate dehydrogenase. Oxaloacetate is regenerated in the cytoplasm by a series of reactions involving malic enzyme, pyruvate carboxylase, and glutamate-aspartate transaminase. The NADPH produced by the malic enzyme reaction is utilized for fatty acid synthesis in the cytoplasm from substrates provided by the glycolytic breakdown of fructose. This mechanism of action did not appear to be important in rat liver cells, since fluoromalate, an inhibitor of malate dehydrogenase, produced only minimal decreases in the fructose-stimulated oxidation of ethanol.[89] In addition, inhibition of glycolysis by iodoacetate or sodium fluoride had no effect on fructose-stimulated ethanol oxidation by rat liver cells.

151

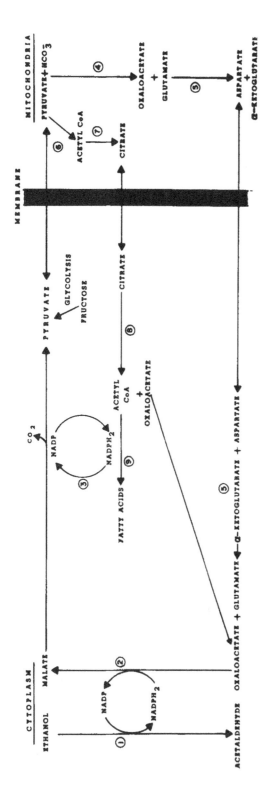

FIGURE 5. Metabolic sequences proposed to explain increases in ethanol oxidation by fructose in perfused pig liver. (1) Alcohol dehydrogenase, (2) malate dehydrogenase, (3) malic enzyme, (4) pyruvate carboxylase, (5) glutamate-aspartate transaminase, (6) pyruvate dehydrogenase, (7) citrate synthetase, (8) citrate lyase, (9) fatty acid synthesis complex. (Adapted from Damgaard, S. E., Sestoft, L., Lundquist, F., and Tygstrup, N., *Acta Med. Scand. Suppl.*, 542, 131, 1972.)

From studies utilizing perfused rat liver it has been proposed that the main site for the final disposal of reducing equivalents produced by the enhanced oxidation of ethanol by fructose is the mitochondrial respiratory chain.[90] In the presence of ethanol, fructose metabolism appears to be shifted away from the energy-producing reactions of glycolysis toward the energy-consuming production of glucose and glycerol. Because the requirements for ATP would be increased under these circumstances, oxidative phosphorylation in the mitochondria would be stimulated with the resultant reoxidation of larger quantities of NADH. This, in turn, would increase the rate of NAD^+-dependent reactions such as that catalyzed by alcohol dehydrogenase. This mechanism is consistent with results from a human study in which the output of lactate and pyruvate by liver was reduced after the infusion of ethanol and fructose as compared to ethanol alone.[92] In addition, glucose output was increased by a factor of three and oxygen uptake by the liver increased by 60% with fructose and ethanol as compared to fructose or ethanol alone.

V. CONCLUSIONS

In this chapter the interactions between fructose and some other components usually present in diets consumed by humans have been described; other examples of such interactions have been alluded to in other chapters. The interaction between the fructose moiety of sucrose and saturated fat resulting in elevated levels of blood triglycerides has been described in Chapter 7. An interaction between fructose and elevated levels of blood glucose achieved postprandially that produces an increased secretion of insulin has been suggested by the material presented in Chapter 5. It is apparent that a multitude of opportunities exist for such interactions between dietary components and that these interactions can have profound metabolic effects quite different from those produced by the dietary components alone. An understanding of the nature of these interactions and their resultant effects on metabolic and physiological processes is an area of nutrition research that merits increased emphasis.

REFERENCES

1. **Macdonald, I.**, Interrelationship between the influences of dietary carbohydrates and fats on fasting serum lipids, *Am. J. Clin. Nutr.*, 20, 345, 1967.
2. **Little, J. A., Birchwood, B. L., Simmons, D. A., Antar, M. A., Kallos, A., Buckley, G. C., and Csima, A.**, Interrelationship between the kinds of dietary carbohydrate and fat in hyperlipoproteinemic patients. I. Sucrose and starch with polyunsaturated fat, *Atherosclerosis*, 11, 173, 1970.
3. **Antar, M. A., Little, J. A., Lucas, C., Buckley, G. C., and Csima, A.**, Interrelationship between the kinds of dietary carbohydrates and fat in hyperlipoproteinemic patients. III. Synergistic effects of sucrose and animal fat on serum lipids, *Atherosclerosis*, 11, 191, 1970.
4. **Mann, J. I., Watermeyer, G. S., Manning, E. B., Randles, J., and Truswell, A. S.**, Effects on serum lipids of different dietary fats associated with a high sucrose diet, *Clin. Sci.*, 44, 601, 1973.
5. **Klevay, L. M.**, The role of copper, zinc, and other chemical elements in ischemic heart disease, in *Metabolism of Trace Metals in Man*, Vol. 1, Rennert, O. W. and Chan, W.-Y., Eds., CRC Press, Boca Raton, Fla., 1984, 129.
6. **Klevay, L. M., Inman, L., Johnson, L. K., Lawler, M., Mahalko, J. R., Milne, D. B., Lukaski, H. C., Bolonchuk, W., and Sandstead, H. H.**, Increased cholesterol in plasma in a young man during experimental copper depletion, *Metabolism*, 33, 1112, 1984.
7. **Klevay, L. M.**, Coronary heart disease: the zinc/copper hypothesis, *Am. J. Clin. Nutr.*, 28, 764, 1975.
8. **Klevay, L. M.**, Interactions of copper and zinc on the occurrence of ischemic heart disease, *Ann. N.Y. Acad. Sci.*, 355, 140, 1980.
9. **Fields, M., Ferretti, R. J., Smith, J. C., Jr., and Reiser, S.**, Effect of copper deficiency on metabolsim and mortality in rats fed sucrose or starch diets, *J. Nutr.*, 113, 1335, 1983.

10. **Reiser, S., Ferretti, R. J., Fields, M., and Smith, J. C., Jr.,** Role of dietary fructose in the enhancement of mortality and biochemical changes associated with copper deficiency in rats, *Am. J. Clin. Nutr.*, 38, 214, 1983.

11. **Fields, M., Ferretti, R. J., Smith, J. C., Jr., and Reiser, S.,** The interaction of type of dietary carbohydrates with copper deficiency, *Am. J. Clin. Nutr.*, 39, 289, 1984.

12. **Fields, M., Ferretti, R. J., Smith, J. C., Jr., and Reiser, S.,** Impairment of glucose tolerance in copper-deficient rats: dependency on the type of dietary carbohydrate, *J. Nutr.*, 114, 393, 1984.

13. **Fields, M., Ferretti, R. J., Smith, J. C., Jr., and Reiser, S.,** Effect of dietary carbohydrates and copper status on blood pressure of rats, *Life Sci.*, 34, 763, 1984.

14. **Fields, M., Ferretti, R. J., Reiser, S., and Smith, J. C., Jr.,** The severity of copper deficiency in rats is determined by the type of dietary carbohydrate, *Proc. Soc. Exp. Biol. Med.*, 175, 530, 1984.

15. **Fields, M., Ferretti, R. J., Smith, J. C., and Reiser, S.,** Interactions between dietary carbohydrate and copper nutriture on lipid peroxidation in rat tissues, *Biol. Trace Elem. Res.*, 6, 379, 1984.

16. **Fields, M., Ferretti, R. J., Judge, J. M., Smith, J. C., and Reiser, S.,** Effects of dietary carbohydrates on hepatic enzymes of copper-deficient rats, *Proc. Soc. Exp. Biol. Med.*, 178, 362, 1985.

17. **Fields, M., Holbrook, J., Scholfield, D., Smith, J. C., Jr., Reiser, S., and Los Alamos Medical Research Group,** Effect of fructose or starch on copper-67 absorption and excretion by the rat, *J. Nutr.*, 116, 625, 1986.

18. **Fields, M., Holbrook, J., Scholfield, D., Rose, A., Smith, J. C., and Reiser, S.,** Development of copper deficiency in rats fed fructose or starch: weekly measurements of copper indices in blood, *Proc. Soc. Exp. Biol. Med.*, 181, 120, 1986.

19. **Jones, J. H. and Foster, C.,** A salt mixture for use with basal diets either low or high in phosphorus, *J. Nutr.*, 24, 245, 1942.

20. **Bieri, J. G., Stoewsand, G. S., Briggs, G. M., Phillips, R. W., Woodard, J. C., and Knapka, J. J.,** Report on the American Institute of Nutrition ad hoc committee on standards for nutritional studies, *J. Nutr.*, 107, 1340, 1977.

21. **Allen, K. G. D. and Klevay, L. M.,** Cholesterolemic and cardiovascular abnormalities in rats caused by copper deficiency, *Atherosclerosis*, 29, 81, 1978.

22. **Viestenz, K. E. and Klevay, L. M.,** A randomized trial of copper therapy in rats with electrocardiographic abnormalities due to copper deficiency, *Am. J. Clin. Nutr.*, 35, 258, 1982.

23. **Cohen, A. M., Teitelbaum, A., Miller, E., Ben-Tor, V., Hirt, R., and Fields, M.,** The effect of copper on carbohydrate metabolism in rats, *Isr. J. Med. Sci.*, 18, 840, 1982.

24. **Owen, C. A., Jr.,** Acquired copper deficiency. Copper deficiency and toxicity, in *Copper in Biology and Medicine Series*, Noyes Publications, Park Ridge, N.J., 1982, 1.

25. **McMurray, C. H.,** Copper deficiency in ruminants. Biological roles of copper, in *Ciba Foundation Symposium 79*, Excerpta Medica, Amsterdam, 1980, 183.

26. **Kellogg, E. W., III, and Fridovich, I.,** Liposome oxidation and erythrocyte lysis by enzymatically generated superoxide and hydrogen peroxide, *J. Biol. Chem.*, 252, 6721, 1977.

27. **Paynter, D. I.,** The role of dietary copper, manganese, selenium, and vitamin E in lipid peroxidation in tissues of the rat, *Biol. Trace Elem. Res.*, 2, 121, 1980.

28. **Balevska, P. S., Russanov, E. M., and Kassabova, T. A.,** Studies on lipid peroxidation in rat liver by copper deficiency, *Int. J. Biochem.*, 13, 489, 1981.

29. **Herndon, J. F., Rice, E. G., Tucker, R. G., Van Loon, E. J., and Greenberg, S. M.,** Iron absorption and metabolism. III. The enhancement of iron absorption in rats by D-sorbitol, *J. Nutr.*, 64, 615, 1958.

30. **Lengemann, F. W. and Comar, C. L.,** Distribution of absorbed strontium-85 and calcium-45 as influenced by lactose, *Am. J. Physiol.*, 200, 1051, 1961.

31. **Bushnell, P. J. and DeLuca, H. F.,** The effects of lactose on the absorption and retention of dietary lead, *J. Nutr.*, 113, 365, 1983.

32. **Favus, M. J. and Angeid-Backman, E.,** Effects of lactose on calcium absorption and secretion by rat ileum, *Am. J. Physiol.*, 246, G281, 1984.

33. **Landes, D. R.,** Influence of dietary carbohydrate on copper, iron and zinc status of the rat, *Proc. Soc. Exp. Biol. Med.*, 150, 686, 1975.

34. **Reiser, S., Bickard, M. C., Hallfrisch, J., Michaelis, O. E., IV, and Prather, E. S.,** Blood lipids and their distribution in lipoproteins in hyperinsulinemic subjects fed three different levels of sucrose, *J. Nutr.*, 111, 1045, 1981.

35. **Woteki, C. E., Welsh, S. O., Raper, W., and Marston, R. M.,** Recent trends and levels of dietary sugars and other caloric sweeteners, in *Metabolic Effects of Utilizable Carbohydrates*, Reiser, S., Ed., Marcel Dekker, New York, 1982, 1.

36. U.S. Department of Agriculture, Sugar and Sweetener Outlook and Situation Report No. SSRV10N1, Economic Research Service, U.S. Government Printing Office, Washington, D.C., 1985, 31.

37. National Academy of Sciences, Recommended Dietary Allowances, 9th Ed., Committee on Dietary Allowances, Food and Nutrition Board, National Research Council, Washington, D.C., 1980.

38. **Holden, J. M., Wolf, W. F., and Mertz, W.**, Zinc and copper in self-selected diets, *J. Am. Diet. Assoc.*, 75, 23, 1979.

39. **Klevay, L. M., Reck, S., and Barcome, D. F.**, Evidence of dietary copper and zinc deficiencies, *JAMA*, 241, 1916, 1979.

40. **Reiser, S., Smith, J. C., Jr., Mertz, W., Holbrook, J. T., Scholfield, D. J., Powell, A. S., Canfield, W. K., and Canary, J. J.**, Indices of copper status in humans consuming a typical American diet containing either fructose or starch, *Am. J. Clin. Nutr.*, 42, 242, 1985.

41. **Charley, P. J., Bibudhendra, S., Stitt, C. F., and Saltman, P.**, Chelation of iron by sugars, *Biochim. Biophys. Acta*, 69, 313, 1963.

42. **Davis, P. S. and Deller, D. J.**, Prediction and demonstration of iron chelating ability of sugars, *Nature*, 212, 404, 1966.

43. **Bates, G. W., Boyer, J., Hegenauer, J. C., and Saltman, P.**, Facilitation of iron absorption of ferric-fructose, *Am. J. Clin. Nutr.*, 25, 983, 1972.

44. Keshan Disease Research Group, Observation on effect of sodium selenite in prevention of Keshan Disease, *Chin. Med. J.*, 92, 471, 1979.

45. Keshan Disease Research Group, Epidemiologic studies on the etiologic relation of Se and Keshan disease, *Chin. Med. J.*, 92, 477, 1979.

46. **Salonen, J. T., Alfthan, G., Huttunen, J. K., Pikkarainen, J., and Puska, P.**, Association between cardiovascular death and myocardial infarction and serum selenium in a matched-pair longitudinal study, *Lancet*, 2, 1975, 1982.

47. **Shamberger, R. J. and Willis, C. E.**, Selenium distribution and human cancer mortality, *Crit, Rev. Clin. Lab. Sci.*, 2, 211, 1971.

48. **Schrauzer, G. N.**, Selenium and cancer; a review, *Bioinorg. Chem.*, 5, 275, 1976.

49. **Griffin, A. C.**, Role of Se in the chemoprevention of cancer, *Adv. Cancer Res.*, 29, 419, 1979.

50. **Jenkinson, S. G., Lawrence, R. A., Burk, R. F., and Williams, D. M.**, Effects of copper deficiency on the activity of the selenoenzyme glutathione peroxidase and on excretion and tissue retention of $^{75}SeO_3$, *J. Nutr.*, 112, 197, 1982.

51. **Levander, O. A., DeLoach, D. P., Morris, V. C., and Moser, P. M.**, Platelet glutathione peroxidase activity as an index of selenium status in rats, *J. Nutr.*, 113, 55, 1983.

52. **Paynter, D. I. and Martin, G. B.**, Investigations into combined dietary deficiencies of copper, selenium and vitamin E in the rat, *Biol. Trace Elem. Res.*, 2, 175, 1980.

53. **Fridovich, I.**, Oxygen radicals, hydrogen peroxide and oxygen toxicity, in *Free Radicals in Biology*, Vol. 1, Pryor, W. A., Ed., Academic Press, Orlando, Fla., 1976, 239.

54. **Wong, L., Deagen, J. T., and Whanger, P. D.**, Effects of dietary carbohydrates on selenium metabolism and glutathione peroxidase activity in rats, *Fed. Proc., Fed. Am. Soc. Exp. Biol.*, 44, 1509, 1985.

55. **Reiser, S. and Christiansen, P. A.**, The properties of the preferential uptake of L-leucine by isolated intestinal epithelial cells, *Biochim. Biophys. Acta*, 225, 123, 1971.

56. **Reiser, S., Michaelis, O. E., IV, and Hallfrisch, J.**, Effects of sugars on leucine and lysine uptake by intestinal cells from rats fed sucrose and stock diets, *Proc. Soc. Exp. Biol. Med.*, 150, 110, 1975.

57. **Reiser, S. and Hallfrisch, J.**, Stimulation of neutral amino acid transport by fructose in epithelial cells isolated from rat intestine, *J. Nutr.*, 107, 767, 1977.

58. **Reiser, S. and Christiansen, P. A.** Intestinal transport of amino acids as affected by sugars, *Am. J. Physiol.*, 216, 915, 1969.

59. **Alvarado, F.**, Amino acid transport in hamster small intestine: site of inhibition by galactose, *Nature*, 219, 276, 1968.

60. **Schultz, S. G. and Curran, P. F.**, Coupled transport of sodium and organic solutes, *Physiol. Rev.*, 50, 637, 1970.

61. **Reiser, S. and Christiansen, P. A.**, The properties of Na$^+$-dependent and Na$^+$-independent lysine uptake by isolated intestinal epithelial cells, *Biochim. Biophys. Acta*, 307, 212, 1973.

62. **Reiser, S., Scholfield, D., Trout, D., Wilson, A., and Aparicio, P.**, Effect of glucose and fructose on the absorption of leucine in humans, *Nutr. Rep. Int.*, 30, 151, 1984.

63. **Elias, E., Gibson, G. J., Greenwood, L. F., Hunt, J. N., and Tripp, J. H.**, The slowing of gastric emptying by monosaccharides and disaccharides in test meals, *J. Physiol. (London)*, 194, 317, 1968.

64. **MacGregor, I. L., Guelles, R., Walts, H. D., and Meyer, J. H.**, The effect of acute hyperglycemia on gastric emptying in man, *Gastroenterology*, 70, 190, 1976.

65. Accident Facts, National Safety Council, Chicago, 1982.

66. U.S. Bureau of the Census, Statistical Abstract of the United States, 1982—1983, 103rd ed., U.S. Government Printing Office, Washington, D.C., 1982.

67. **Merry, J. and Marks, V.**, Effect on performance of reducing blood-alcohol with oral fructose, *Lancet*, 2, 1328, 1967.

68. **Pawan, G. L. S.**, Vitamins, sugars and ethanol metabolism in man, *Nature*, 220, 374, 1968.

69. **Wallgren, H. and Barry, H.**, in *Actions of Alcohol*, Vol. 1, Elsevier, London, 1970, 275.

70. **Hoensch, H. and Demling, L.**, The influence of exogenous substrates on the breakdown of alcohol, *Acta Hepatogastroenterol.*, 19, 401, 1972.
71. **Clark, E. R., Hughes, I. E., and Letley, E.**, The effects of oral administration of various sugars on blood ethanol concentrations in man, *J. Pharm. Pharmacol.*, 25, 319, 1973.
72. **Soterakis, J. and Iber, F. L.**, Increased rate of alcohol removal from blood with oral fructose and sucrose, *Am. J. Clin. Nutr.*, 28, 254, 1975.
73. **Carpenter, T. M. and Lee, R. C.**, The effect of fructose on the metabolism of ethyl alcohol in man, *J. Pharmacol. Exp. Ther.*, 60, 286, 1937.
74. **Carpenter, T. M. and Lee, R. C.**, The effect of glucose on the metabolism of ethyl alcohol in man, *J. Pharmacol. Exp. Ther.*, 60, 264, 1937.
75. **Stuhlfauth, K. and Neumaier, H.**, Die Wirkung der Laevulose auf Alkohol-Intoxikationen, *Med. Klin.*, 46, 591, 1951.
76. **Pletscher, A., Bernstein, A., and Staub, H.**, The acceleration of alcohol catabolism by fructose in man, *Experientia*, 8, 307, 1952.
77. **Lundquist, F. and Wolthers, H.**, The influence of fructose on the kinetics of alcohol elimination in man, *Acta Pharmacol. Toxicol.*, 14, 290, 1958.
78. **Patel, A. R., Paton, A. M., Rowan, T., Lawson, D. H., and Linton, A. L.**, Clinical studies on the effect of laevulose on the rate of metabolism of ethyl alcohol, *Scott. Med. J.*, 14, 268, 1969.
79. **Meyer, B. H., Muller, F. O., and Hundt, H. K. L.**, The effect of fructose on blood alcohol levels in man, *S. Afr. Med. J.*, 62, 719, 1982.
80. **Rawat, A. K.**, Effects of fructose and other substances on ethanol and acetaldehyde in man, *Res. Commun. Chem. Pathol. Pharmacol.*, 16, 281, 1977.
81. **Lowenstein, L. M., Simone, R., Boulter, P., and Nathan, P.**, Effect of fructose on alcohol concentrations in the blood of man, *JAMA*, 213, 1899, 1970.
82. **Brown, S. S., Forrest, J. A. H., and Roscoe, P.**, A controlled trial of fructose in the treatment of acute alcoholic intoxication, *Lancet*, 2, 898, 1972.
83. **Bode, C., Bode, C., and Thiele, D.**, Alcohol metabolism in man: effect of intravenous fructose infusion on blood ethanol elimination rate following stimulation by phenobarbitol treatment or chronic alcohol consumption, *Klin. Wochenschr.*, 57, 125, 1979.
84. **Sprandel, U., Troger, H.-D., Liebhardt, E. W., and Zollner, N.**, Acceleration of ethanol elimination with fructose in man, *Nutr. Metab.*, 24, 324, 1980.
85. **Camps, F. E. and Robinson, A. E.**, Influence of fructose on blood alcohol levels in social drinkers, *Med. Sci. Law*, 8, 161, 1968.
86. **Franks, H. M., Hagedorn, H., Hensley, V. R., Hensley, W. J., and Starmer, G. A.**, The interaction of fructose, dextrose and ethanol on human performance, *Clin. Exp. Pharmacol. Physiol.*, 4, 153, 1977.
87. **Ylikahri, R. H., Leino, T., Huttunen, M. O., Poso, A. R., Eriksson, C. J. P., and Nikkilä, E. A.**, Effects of fructose and glucose on ethanol-induced metabolic changes and on the intensity of alcohol intoxication and hangover, *Eur. J. Clin. Invest.*, 6, 93, 1976.
88. **Lundquist, F.**, The effect of ethanol on liver metabolism, in *Alcohol and Alcoholism*, Pophorn, R. E., Ed., University of Toronto Press, Canada, 1970, 33.
89. **Berry, M. N. and Kun, E.**, Ethanol oxidation by isolated rat-liver cells, *Eur. J. Biochem.*, 89, 237, 1978.
90. **Scholz, R. and Nohl, H.**, Mechanism of the stimulatory effects of fructose on ethanol oxidation in perfused rat liver, *Eur. J. Biochem.*, 63, 449, 1976.
91. **Damgaard, S. E., Sestoft, L., Lundquist, F., and Tygstrup, N.**, The interrelationship between fructose and ethanol metabolism in the isolated perfused pig liver, *Acta Med. Scand.*, Suppl. 542, 131, 1972.
92. **Tygstrup, N., Winkler, K., and Lundquist, F.**, The mechanism of the fructose effect on the ethanol metabolism of the human liver, *J. Clin. Invest.*, 44, 817, 1965.

INDEX

C

D

H

Printed and bound by CPI Group (UK) Ltd, Croydon, CR0 4YY

22/10/2024

01777600-0016